2021

Dean
Kamaye

2021

A RUNNER'S
HIGH

A RUNNER'S
HIGH

MY LIFE IN MOTION

DEAN KARNAZES

HarperOne
An Imprint of HarperCollinsPublishers

HarperCollins books may be purchased for educational, business, or sales promotional use. For information, please email the Special Markets Department at SPsales@harpercollins.com.

FIRST EDITION

Library of Congress Cataloging-in-Publication Data

Names: Karnazes, Dean, 1962- author.
Title: A runner's high / Dean Karnazes.
Description: First edition. | New York : HarperOne, 2021.
Identifiers: LCCN 2020039168 (print) | LCCN 2020039169 (ebook) |
    ISBN 9780062955500 (hardcover) | ISBN 9780062955517 (trade paperback) |
    ISBN 9780062955555 (ebook)
Subjects: LCSH: Karnazes, Dean, 1962- | Running—Psychological aspects.
    | Ultra running. | Runners (Sports)—United States—Biography.
Classification: LCC GV1061.15.K39 A3 2021 (print) | LCC GV1061.15.K39 (ebook)
    | DDC 796.42092 [B]—dc23
LC record available at https://lccn.loc.gov/2020039168
LC ebook record available at https://lccn.loc.gov/2020039169

21 22 23 24 25   LSC   10 9 8 7 6 5 4 3 2 1

This book is dedicated to my mother and father—
to the times we've had, the memories we've created,
and the joy we've shared. It's been a fantastic journey.
Together forever.

CONTENTS

A RUNNER'S
HIGH

1

ENDURANCE NEVER SLEEPS

Running an ultra is simple;
all you have to do is not stop.

I'm lying catawampus splayed ass-to-the-dirt in the trail—one leg tweaked improbably beneath me—staring up at the afternoon sky seeing sparkles of light flickering before me like circling fireflies and wondering what the hell just happened. A sharp ringing in my ears perforates the otherwise complete stillness, a lazy film of dust rising indolently around my idle carcass. Inside the motor room my muscles and bodily organs register a dull tenderness, but it is the nausea that is most pronounced, a queasy sensation of being punched hard in the gut. What just happened?

Moments ago I was in perfect harmonic flow, bounding along nicely, cool and in control, step, spring, step . . . Then everything changed. I vaguely recall flight, weightless soaring, a defiant middle finger to gravity as time briefly suspended; my wings spread—fly, be free . . .

Until impact. Kaboom! Everything just exploded, like a sky-diver whose chute failed to deploy. Now I'm heaped on the soil like Icarus, a lifeless, charred exoskeleton smoldering in ruin and wondering what just went down. A ticker tape of questions scroll across the screen of my mind: Is anything broken? Will someone find me? Where am I?

To answer that final question we need to dial back the clock to yesterday morning, a time when I had a sinking premonition: *I shouldn't be doing this. I REALLY shouldn't be doing this. I know better.* Then I shut the door behind me. I was doing it.

At least the timing of my departure seemed good. The merciless Bay Area traffic was showing its gentler side and I slipped through the busiest corridors with barely a tap on the brakes. Sometimes it takes hours just getting across town, and when it comes to sucking the living soul out of a creature, perhaps no human creation is more noxious than traffic (with the exception of TSA lines).

Still, despite the absence of congestion, it took nearly eight hours to reach my destination, the juxtaposed pastoral hamlet of Bishop, California. Nestled under the striking peaks of the Eastern Sierra Nevada mountain range, Bishop is something of a conundrum. It's in a beautiful natural setting, though one oddly frequented equally by hikers and bikers (and the bikes they're riding aren't the kind with pedals). The main street through town has quaint galleries, outdoor mountaineering stores, a nature center, and an indie bookstore, things you might expect in a mountain settlement. But then there are rows of fast-food joints, seedy bars, a collection of budget hotels, and a Kmart, all of which thoroughly taper the city's charm with a liberal dousing of contemptible.

I was meeting my father here, at one such establishment of lesser repute. Unfortunately, there was little choice in the matter; it was the only remaining hotel room in town. Reservations were made last minute and I booked what I could get. As would be expected on such short notice, there also weren't many options for securing a crew to help support my endeavor, though I somehow snagged the very best (i.e., dear ol' Dad). Who else would drop everything on a single two-minute phone call and drive six hours from Southern California to meet me? There hasn't been a more loyal companion in my life than my father.

A spry eighty-two years old, the man bounced about like a loosely attached valence electron careening haphazardly around its outer shell. Sparks flew off him, a perpetual fission reaction capable of erupting with no forewarning. He was electric, charismatic, overwhelming at times, and wholly uncontainable. Every moment with him was slightly unpredictable. The older he grew the more lively his personality became. Laughter, angst, melancholy, joy—all of these emotions could be expressed within the confines of a single brief interaction. You never knew what to expect with Dad.

"ULTRAMARATHON MAN!" he boomed when he saw me (I'd asked him not to call me that a thousand times, but it was no use). A reporter had tagged me with that lovely moniker and I'd never felt comfortable with it. But over time it had taken on a life of its own, especially with my dad!

"Hiya, Pops," I said, hugging him. "How was the drive?"

"Piece of cake." He was fond of clichés.

"So you good?" I asked.

"Never had a bad day."

Just wait until tomorrow, I thought coyly.

My mother was usually part of these far-flung escapades. The two were nearly inseparable. Sixty years of wedlock had brought them closer, two old-fashioned romantics clinging tightly through all of life's crazy turbulence. Since their retirement they were in a state of perpetual motion. They'd toured just about every part of North America, Australia, and much of Europe. Sometimes on a whim they'd fly to Greece for a month or two with no fixed plans, no itinerary, no accommodations, nothing but a rental car (and rental cars in Greece are not always the most dependable machines). "Things work out," my mother always tells me. She wasn't here today because of a 5K she'd scheduled along the beach with her buddies, most of whom were decades younger. They still couldn't keep up. She wasn't fast, but my mother had the gift of endurance. From the Greek island of Ikaria—one of the fabled "Blue Zones" where indigenous people routinely live beyond a hundred—she is freakishly indefatigable, especially when it comes to outdoor adventures. Mom would certainly be with us today if she weren't showing up those young lasses back home.

The air in Bishop is different than in San Francisco. In the Bay Area, even when you can't see the water you can still smell its thick, salty dampness. In Bishop the air is hot and arid, a subtle hint of a smoldering campfire permanently hangs in the atmosphere. You could feel it in your eyes, the gritty dryness, and in your sinuses. Bishop sits in the high desert, in the lee of an imposing mountain range. Incoming storms lose their moisture as they sweep across California, and any rainfall left as they progress inland is mostly deposited along the western slopes of the uprising. Perilously little water makes it over the towering granite impediment of the Sierra Nevada. On average, Bishop receives about five inches of rainfall annually, and summertime humidity can drop into the single digits. Think of it as having a hair dryer ceaselessly blowing instead of an encroaching fog bank.

Although now well into the afternoon, the sun still seared my skin as I walked to the office to collect our room keys. The official start to summer wasn't for a few weeks, but you'd never know it. The heat coming off the pavement radiated through my shoes, warming and swelling my feet. Tomorrow was supposed to be even hotter.

A small wall-mounted air-conditioning unit was noisily sputtering in the corner when I entered, but it was simply no match for the elements. It was stifling inside, even though the shades were drawn and it was dark. The innkeeper used a handkerchief to pat the sweat off his forehead. The place reeked of Lysol and dirty socks. I asked if there was an ice machine. "There is," he told me. "But it's busted."

The elevator was busted, too. Thus we carried our bags up to our second-story *chalet*. "Sorry about these accommodations."

"They're fine," my dad offered up, "just fine."

Staying in the room next to us were two fully grown pit bulls. I was told the hotel was "pet-friendly," but two adult pit bulls hardly seemed like sociable pets to me. The owners didn't appear very genial, either. Standing outside having a cigarette, they looked us over with suspicion.

And we, for our part, were quick to get in our room and shut the door behind us. Once inside, the place was musty and dank. "We should probably check for bedbugs," I bemoaned, hoisting our bags into the closet. But when I pulled open the blinds to let in some light the view out the dusty window instantly carried me someplace else, someplace special and expansive, a familial place that was part of my very constitution. Beams of late-afternoon sunlight extended heavenward, the jagged silhouette of the Sierra Nevada perched in the distance like an Ansel Adams photograph, towering columns of marble white clouds rising into the air and the sky so impossibly deep, dark blue. I'd been coming here most my life, since Dad and I first climbed Mount Whitney—the highest peak in the conterminous United States—when I was twelve years old. We carried heavy metal-framed packs and slept in a thick canvas tent, our hiking boots and wool socks left outside to air out. We cooked pouches of freeze-dried food over a small camp stove and rationed water from our canteens until we could find another brook to refill them. During the day we hiked, eating leathery beef jerky and trail mix, my fingers colorfully dyed with the melted coating of M&Ms. Sometimes we talked, but mostly we just hiked, swept up in the grand enormity of the surroundings, the brilliant artistry of Mother Nature holding us spellbound. When we reached the summit I humbly signed the logbook, forever marking my presence on this hallowed mountain peak.

I wasn't a very good student, but my writing assignment about the Eastern Sierra trip with Dad got an A plus. It was my first A plus ever, and the teacher had plastered the report with a bunch of those colorful smiley-face stickers. They were stuck all over it, colorful little dots, and it made me joyful seeing all those smiley faces, a warm, flush feeling inside.

I loved those days, and I loved those adventures. I could let my long, wavy hair go uncombed. Nobody told me I couldn't go there or I couldn't do that; out here I was the master of my own destiny, free to wander as I pleased, free to explore. We didn't have much when I was a boy, we had everything. We had the Eastern Sierra, Yosemite, and Sequoia. We had the San Gabriels and San Jacinto. We had Joshua Tree and Death Valley, Tahoe and Desolation Wilderness. We had Big Sur and the Pinnacles, Mendocino and the Redwoods in the north, Shasta in the middle, and Lassen to the east. We had California wild and untamed, and every summer vacation, every spring break, every school holiday and long weekend, we'd pack up the lime-green Ford Country Squire station wagon (replete with wood panels) and head for the trails. We were *Outside* magazine before there was *Outside* magazine.

Tomorrow I'd be setting out to relive some of those memories and to create some new ones. I'd returned to run the Bishop High Sierra Ultramarathon and my dad and I were together once again, a team reunited. Older now, yes, but still together. Still carrying on.

The Bishop High Sierra Ultramarathon offered four race distances: 20 miles, 50 kilometers, 50 miles, and 100 kilometers. "I'm not in any shape to run 100 kilometers," I told Dad.

"So which race did you sign up for?" he asked.

"The 100 kilometers."

Of course I did. "I shouldn't be doing this," I said to him. "I know better."

"This isn't your first rodeo, cowboy."

"Yeah, I guess you're right. I've done some stupid shit before."

"C'mon, ultramarathon man, you know what you're up against," he said, patting me on the back.

"Yeah, I do know what I'm up against. And that's what scares me."

What I was up against was 62 miles of climbing and descending a narrow dirt path through the mountains and desert of the High Sierra in the blazing heat. I knew full well what I was up against. But another battle awaited me before I even reached the starting line.

"I'll set the alarm for three thirty."

"Three thirty! Why so early? The race doesn't start till five thirty."

"You don't want to be late."

"Dad, it's a five-minute drive."

"You want to have time to warm up?"

"Warm up? I'll have 62 miles to warm up."

"Suppose there's traffic?"

"Dad, this is Bishop, population 3,760. The only way there'd be traffic is if there's an earthquake."

"Suppose there is?"

"AHH! You're exhausting."

Arguing with Dad could be more draining than running an ultramarathon. One of his most vigorously defended positions has to do with promptness. In my opinion he takes it too far. For instance, if he has an appointment scheduled—say, at the DMV—he would make it a point to arrive at least an hour early, just to be on the safe side. Now, I'm not sure about you, but if I found myself with a spare hour to burn, waiting at the DMV wouldn't top my priority list. But there was no use arguing with the man.

"Okay, Pops, set your alarm for three thirty."

"Great. We'll have some coffee."

We both looked up at the chintzy in-room coffeemaker. There were two Styrofoam cups, a generic Mylar bag stamped *Supreme Coffee*, and a couple of pink packets of artificial sweetener. Yum.

"Do you see a socket over there?" he inquired.

I looked behind the table separating our beds. "There's one right here."

"Could you plug this in?" He handed me the extension cord from his CPAP machine.

"You're gonna wear that thing tonight?"

"Doc says I should use it every night." The CPAP is an apparatus used to treat sleep apnea. Continuous positive airway pressure (CPAP) uses a hose and head mask to enhance nighttime breathing, and there's a small water chamber that adds moisture to the airflow. When he had it on he looked like Hannibal Lecter and sounded like a deep-sea diver. All night long this rhythmic gurgling gently serenaded me, like a lapping tide ebbing and flowing. It was as though I were sleeping on a dock.

The chiming of the three thirty alarm was inconsequential. I hadn't slept much all night, not with Aquaman bubbling away next to me. But that was fine; endurance never sleeps.

Splashing some water on my face over the bathroom sink, I peered at myself through the yellowy light. *I shouldn't be doing this*, I thought. I haven't been training; I know better. I rubbed my fingers through my hair. *Let's do this.*

There wasn't much activity when we arrived at the starting area, which was a large open park with a lake in the middle, the reflection of the moon rippling on its surface. Gradually dawn began to emerge and a crowd slowly gathered. All of the race distances—20 miles, 50K, 50 miles, and 100K—started at the same time, and more and more racers began appearing. In the assemblage I saw a few familiar faces, Billy Yang being one of them.

"Karno! How's it going, bro?"

"Hey, Billy, nice to see you."

"You're the reason I'm here, dude. Don't you ever forget that."

"I'm just glad we're still friends."

He laughed. Billy credited me as one of the forces behind his journey into ultramarathoning. He'd read one of my earlier books and decided to give it a go. A fresh young face in the mix, it was good to see him.

"You running the 100K?" I asked.

"No way, dude. The 50K. I haven't been training. I know better."

Doh! How someone with less experience than I had could somehow be wiser was a quandary of ongoing internal debate. Consistently biting off more than I could chew and wantonly hurling myself into waters over my head were parts of my primary source code, repeating themes. Perhaps I subconsciously thrilled at the possibility of shit going wrong. Look, I'm just as unstirred by life as the next guy. Modern existence is so comfortably predictable, so ho-hum, so, well, boring. But *this* was different. With an ultramarathon the outcome was never certain, and when things unraveled they truly fell apart. It elevated having a bad day at the office to unprecedented heights. And I lived for it.

The race director stepped to the front of the group to say a few parting words. He warned us to stay hydrated, as it was going to be hot and dry today. He alerted us to watch for trail markers to avoid getting lost. Again he cautioned us to stay hydrated.

There were 181 runners at the start. Old-timers lament about the unfettered growth and mainstream popularity of today's ultrarunning scene. Out of control, they say. The previous year's New York City marathon had more than 55,000 registered participants. I'd say the sport of ultrarunning still had a bit of growth left, not that I'm taking sides in the matter. I see both points. When I first started this crazy sport, if fifty people showed up for a race it was a pretty good turnout. Nowadays some ultramarathons sell out, and there are even lotteries to get in. Blasphemy!

I stepped to the side of the pack to say good-bye to Dad. "Keep your nose clean, kid," he counseled. He'd been giving me this same advice for most my life. I still had no idea what it meant.

"Thanks, Pops. Will do."

The countdown commenced. "*. . . four, three . . .*" I made a cross on my chest. "*. . . two, one . . .*" The starting gun fired and the pack surged forward. Church was now in session.

I established my position squarely in the thick of the lead pack and fixed my gaze on two sets of muscular calves in front of me, running in tight formation with a cluster of others somewhere near the head of the herd. The initial terrain was relatively flat and groomed, so the pace was hasty, the starting line adrenaline still working its way through the system. A spiraling cloud of dust arose in the still morning air like a wildebeest stampede crashing forth with frenzied urgency.

This lasted for all of about 5 miles. Then the incline steepened, and maintaining such a rushed cadence became unsustainable if not impossible. My lungs stretched to full capacity yet still couldn't deliver enough oxygen to propel my legs at that same clip. Hence, I slowed. Over the next 15 miles the course would gain about five thousand feet of elevation. For perspective, the notorious Heartbreak Hill at the Boston Marathon rises less than a hundred feet. An ultramarathon is a different sort of monster.

And ultramarathoners are different sorts of creatures. Instead of seeking flat and fast courses, we look for the hilliest and most challenging ones. Speed matters, but the elevation gain and loss of a particular race are equal badges of honor. The legendary Hardrock 100, for instance, has 33,050 feet of climbing and descending, for a total elevation change of 66,100 feet. That's the equivalent of running from sea level to the summit of Mount Everest and back, plus a warm-up and a cool-down.

Ultramarathoners abide by a different dress code, too (or irreverence for one). Mismatched colors are commonplace. Vibrant crew socks, Day-Glo arm sleeves, funky trucker hats, retro eyewear—nothing is too outlandish. The same goes for tats, hair color, piercings, and beards. They're commonplace. Though as far as I'm concerned, if you've got the guts and the discipline to be running an ultramarathon you can pretty much wear whatever the hell you want.

At the 15-mile mark I was feeling light and nimble. At 20 miles I was feeling light-headed with a slight ring in my ears. I knew this feeling and knew the cause. Twenty-four hours ago I was a sea dweller, living and sleeping in the thickest air possible. Now I was ninety-four hundred feet up in the air. The altitude was doing its thing.

Standing at the Overlook aid station trying to gather my senses, another two runners came bounding in. We exchanged nods, and began foraging for morsels of food spread about the small camping table.

"How you feeling?" the gentleman asked.

He was youthful-looking and solid as oak. There was a woman running with him; she looked equally buff.

"I'm a bit light-headed, honestly. You guys?"

He smiled sheepishly. "We live in Lake Tahoe."

While he didn't answer my question directly, his meaning was implicit. They lived at altitude and they were coping just fine, as breezily as walking to the end of the driveway to fetch the morning paper. Off they went. I capped my water bottle and strode along after them.

The course had followed a rugged jeep road for much of the distance thus far, but now it meandered through high-desert chaparral in a series of broad switchbacks and rolling ascents and descents. The 20-miler and 50K runners had split off from the 50-mile and 100K runners by this point and were now on their return journey while we still had many miles to go. With the separation of the two groups there were noticeably fewer people on the course and I found myself running solo, no fellow humans in sight.

There'd been a few pine trees interspersed at the higher elevations, but here there was absolutely no shade along the course.
I'd opted to go with a handheld water bottle instead of a larger
hydration pack. It was a strategic decision. Running in a slightly
depleted state was part of my master plan to hack my training.

You see, I needed to seriously accelerate my fitness because
of some news I'd recently received. Good news, for sure, but
also worrisome. Worrisome to the point that immediate action
was required, hence the hasty registration for a tough 100K, the
quick drive to Bishop, and the decision to carry only one handheld water bottle (and an uninsulated one at that). I shouldn't be
doing this, but here I was.

And thus I kept doing it. Mile 20 passed. Mile 23 passed. Mile 26
passed. Running an ultra is simple; all you have to do is not stop.

Bishop Creek was the first place I rendezvoused with Dad. It
was mile 29 and by now the sun was perched firmly overhead,
and scorching. Crewing for someone during an ultramarathon
can best be described as rushing around from one place to the
next, only to sit and wait. Some people are better cut out for this
duty than others. Dad was extremely well adept. He could make
friends with anyone, anywhere. I remember a telemarketer once
calling our house trying to sell him life insurance. Dad turned
the conversation around and ended up selling the guy our used
car. And he wrote Dad a thank-you letter. The man was just that
likable.

And did I mention prompt? Yes, I did. And to prove my point,
he'd been waiting at the Bishop Creek aid station for hours, just
rapping out with people and making friends, and trying to stay
in the shade. But even in the shade it was broiling. Still, he had
a beach chair set up, and all my gear spread out on a towel for
easy access.

"Hey, Pops," I said, running into the aid station. "It's hot."

"It's not too bad. Whaddya need?"

"An air conditioner would be nice."

He took off his hat and started fanning me.

Dad would do anything for his family. Never did he complain or act put out by a request. He was always there, always available. Growing up, he attended all of my games, though I can't say I always appreciated his presence. Some of my friends' fathers never came to theirs. I didn't understand the magnitude of Dad's loyalty back then. Now I do. Family matters; family matters more than all else.

"Say Pops, is that ginger around?"

Ginger is the miracle elixir for gut issues, which is a frequent ailment afflicting long-distance runners, especially when temperatures elevate. Many ultramarathoners chew on candied ginger. I preferred my tonic straight.

Dad handed me a container of freshly sliced ginger. I stuck a piece in my mouth and immediately felt the burn. I encourage you to try this at home (over the sink). Few people can deal. Candied ginger is spicy; straight ginger is the equivalent of tossing live firecrackers into your mouth.

I winced.

Pops looked at me concerned. "I don't know how you do that."

"Whatever works, right?"

He shrugged his shoulders and gave me a doubtful look.

We filled my water bottle and sprayed some additional sunscreen on the back of my neck. As I was leaving, another runner came springing in. He looked young and strong. When he saw me he did a double take. Was he competitive? Did I now have a target on my back?

It was hard to know where anyone stood in the race at this point. Of the few runners I periodically saw it wasn't entirely clear if they were running the 50-miler or the 100K; both of us currently ran on the same course. By this point the pack had thinned considerably and there wasn't much passing or being passed going on; everyone had pretty much established his or her position and was holding his or her own. And as helpful as the volunteers were, they didn't have much information about race standings.

I've never been much of a competitive runner. Sure, I enjoy the thrill and high drama of competition, but I don't live for it. To me, running is a grand adventure, an intrepid outward exploration of the landscape and a revealing inward journey of the self. These are the things that keep me going, the lust for exploration and the quest to better comprehend who I am and what I'm made of. In a lifetime's worth of competition, there's only been one race I explicitly set out to win, the 2004 Badwater Ultramarathon (which I was fortunate enough to actually win). Other than that, running and racing has been an experiential trip, not a desire to end up on the podium.

Still, when someone is hunting you down, you don't want to be passed. Consequently, when I left the aid station I kicked up the pace a notch to avoid capture. The next several miles to the Intake aid station were run aggressively, perhaps too aggressively. It was a short distance; that was the rationale behind my recklessness (that, and not being passed). But when I arrived at the Intake aid station—32 miles from the start—I was suddenly feeling the ramifications. I'd crossed the midpoint but was feeling right about done. That warm, rewarding, semifried, halfway-finished feeling was more akin to something fully charred, like lamb smoldering over an open spit. Suddenly it felt hot. Unbearably hot. Roasting hot. I was cooked, yet still had plenty of course yet to cover.

"Dad, who turned up the furnace?"

He gave me a look like *You think it's hot out there running, try standing around this aid station for an hour.* All of the volunteers looked wilted. There was just one small umbrella set up over the food table, and everyone scrunched together trying to fit under the minuscule shadow it cast.

My original goal at this aid station was to relax and gather my wits—I was feeling a bit loopy—but it was clear this wasn't the place to do it. Consequently, my mission shifted from getting to the aid station to getting out of the aid station just as fast as I could, both for me and Dad. We refilled my water bottle and I grabbed a packet of nut butter, then off I set . . . just as that younger runner was pulling in.

The gap between us had narrowed.

2

GROWING PAINS

Until you go over the edge you don't
know how far the edge is.

As my "career" as an ultramarathoner broadened, so, too, did the sport itself. Ultramarathoning has now entered the mainstream lexicon, and it's not uncommon these days to hear references to the sport in the news or read about ultramarathoning in the paper. Along with this increased notoriety, the number of participants has equally expanded. In 1993—the year I ran my first ultra—I was one of 3,754 finishers in North America. By 2019 that number had swollen to more than 127,296 finishers. And in some other countries this growth has been even higher. In the past twenty-three years global participation in the sport has increased by 1,676 percent, according to a report by the International Association of Ultrarunners. Bear in mind that these are still relatively small numbers,* but when you try cramming all these people down a narrow path through the wilderness, it creates challenges. The sport is experiencing growing pains, and the more established and iconic ultramarathons have set up lottery systems to gain entry. The granddaddy of them all, the Western States 100-Mile Endurance Run, now has slimmer odds of getting in than being accepted to Harvard.

Western States was my first 100-miler, and it will always retain its place as my favorite. I first ran the race in 1994. Not surprisingly, I got in on my first attempt (lottery odds back then were about fifty-fifty). I've run it on more than ten occasions since, earning a coveted 1,000 Miles, Ten Days buckle. A priceless medallion these days, not that there aren't other athletes capable of this feat, but gaining entry into Western States ten times requires overcoming nearly impossible odds.

I hadn't run the race in years. It's not that I was burned out, it's just that some of the enchantment had waned. The race was known territory—not stale, but well trodden. I'd now accumulated so many silver buckles—the prize for finishing the race in fewer than twenty-four hours—that there seemed little left to discover. To that point, my last attempt at the Western States 100 had ended dreadfully in a DNF.† In my hundreds of races over multiple decades, I'd dropped out only a handful of times. Western States 2009 was one of them. Sure, my body was wrecked during this race, but my body had been wrecked before and I'd still continued. Something was missing in my heart; my spirit had been broken.

---

* For comparison, 518,916 people finished a marathon in North America in 2019, and 2.4 million completed a half marathon.

† Did Not Finish.

Yet the heart can be a fickle creature. Old flames mysteriously reignite. After being thoroughly extinguished on the Western States trailside for nearly a decade, an internal fire inexplicably reignited. I wanted back in.

Though, I'll be honest, the thought of returning terrified me. The 2009 race had thoroughly kicked my ass; the DNF haunted me to this day. Yet I wanted—needed—redemption.

I'd applied for the 2018 race. Now all I had to do was win the lottery. Which I didn't, though it came as little surprise. After all, as I've just explained, odds of getting in are infinitesimally small. Though I wasn't a complete loser, I was "wait listed" (i.e., one of fifty individuals put on a list should a current entrant withdraw). I was number 23, not encouraging. But withdrawals can happen. For instance, space aliens could abduct someone. Or a giant sinkhole could swallow an entrant. People have been known to spontaneously combust. My point is this: getting into the Western States 100 is so difficult that it would take an act of God for anyone to withdraw. Number twenty-three on the wait list may as well be infinity.

To their credit, the Western States race organizers go to extraordinary efforts to ensure that the lottery is fair. Every year that a runner is not chosen in the lottery his odds are improved. This is done through the allotment of an extra lottery ticket. In December—when the lottery is held—all of these tickets are placed into a "hat" (actually it's a large drum) and randomly drawn by members of the audience at a public forum. For the 2018 race there were 15,074 tickets in the drum for 369 available starting positions. Incredibly slim odds, and for a seemingly arbitrary number of openings, right? Why 369?

This explanation will help illustrate the difficulties of staging an ultramarathon. In 1984, Congress enacted the California

Wilderness Act, which created the Granite Chief Wilderness Area, a mountainous expanse in the Sierra Nevada that the Western States Trail dissects. Organized sports events are normally not permitted in the Granite Chief Wilderness Area. But the Western States race preexisted the Wilderness Act; thus it was "grandfathered" into the legislation, with the stipulation that no more runners could enter the race in future years than had done so prior to the passage of the Act. In 1984 there just happened to be 369 runners. That has been the magic number ever since.

Even being selected on the wait list, with more than 15,000 tickets in the bucket, was extraordinarily lucky. But it hardly seemed to matter. No one was going to withdraw, not after having worked so hard just to gain entrance. I wrote off my Western States entry attempt as having almost defied overwhelming odds. Almost.

But then I got a strange email from someone alerting me that I was moving up on the list. True—by the first week of April, five people had withdrawn, so I'd advanced from number twenty-three on the wait list to number eighteen. In the five months since the lottery was held, five people had withdrawn. That's an average of one person per month. The race was in June, about two months away. Perhaps another two more people would withdraw, so I would be sixteenth on the list, still a long ways off from standing at the starting line.

Yet in the back of my mind I was thinking that maybe I should start training. Still, the rational side of me argued that it would likely be for naught. There were better ways I could be using my time.

I was to learn a valuable lesson. Never listen to your rational mind. With three weeks remaining before the Western States race day, a total of sixteen people had withdrawn. Now I'd moved up to number seven on the list. Oh, shit, maybe I'll get in after all! And three weeks isn't much time to train. My solution: sign up for a tough 100K and drive to Bishop. We runners aren't compulsive, are we? Nah . . .

Illogical, too (at least this runner). Running a 100K with little base training was the equivalent of getting in the ring with George Foreman and praying for a miracle. Hey, they can happen. After all, I could be getting into Western States.

This is why I was here in Bishop today running the Bishop High Sierra 100K. If the miracle came true and I got into Western States, I needed some base mileage on the chassis. But exiting the Intake aid station at mile 32 I was running out of gas. Both my body and my mind were in a bad place, feeling the collective strain of taking on this race in such a woefully undertrained state. I was hurting, no denying that. And some young punk kid was chasing me down. I didn't want this to happen. I didn't want this to happen because there was more at stake here than just race results. Something deeper.

The course climbed over a tall ridge in a series of switchbacks and then dropped into a long, narrow valley. There was no shelter from the sun, and the path underfoot was dusty, my feet kicking up a fine talcum that hung lazily in the air a few feet off the ground. I was leaving a silt tracer, and anyone behind me would know of my recent passing by the powdery cloud dissipating in my wake. And I knew there was someone behind me; I just didn't know how far behind.

Although the heat seemed to magnify between the valley walls, it was mostly a nice, straight descent down the gorge, and the chalky earth underfoot was forgiving. In ultramarathoner-speak, the trail could be referred to as "runable" (as opposed to unrunable), and I managed to hold my own. I wouldn't say I was flying down the trail, but I was keeping a steady pace and clicking off some of the easier miles. Mind you, the path underfoot was merciful, so what I was doing wasn't all that difficult.

Unfortunately, we didn't ride out this valley to its terminus. Instead, the course banked a sudden left-hand turn and headed up the northern valley wall. The ascent was sheer and abrupt (i.e., unrunable). I power hiked as fast as I could. But the grade steepened, causing my pace to falter. I put my hands on my quadriceps and pushed down with every stride, trying to get a bit more oomph out of each footfall. It was laboriously slow going. At points I had to stop entirely to gather my breath. We were eighty-five hundred feet above sea level; my swollen hands were testament to that. I continued powering onward, doing the best I could.

Suddenly the hairs on the back of my neck prickled. Someone was close behind. I didn't turn to look, but I could hear them, their footsteps, their deep, heavy breathing. I kept my head forward and picked up the pace, hoping, praying that I could somehow drop them and reclaim a respectable measure of distance between us. But they stuck to me, matching my hastened steps with powerful strides of their own. And then came the words, those horrible, cutting words that wounded like a kick to the abdomen, "On your left."

It was that young, fit kid, and he was going to pass me. As he did so, I could sense that he was far less strained than I was; he looked almost comfortable in his brisk ascent of the switchbacks. And there, on that magnificent and desolate stretch of trail in the great Sierra Nevada mountain range, my existential nightmare was affirmed. I didn't so much fear losing to a younger kid; what I feared was losing my relevance. In a sport that had become my life, I didn't want to fade away. I didn't want to become some faceless nobody that gets left in the dirt midway through a race.

Permit me to digress and explain something else about the Western States 100. There are other ways than the lottery to gain entry into the race. If you finished in the top ten one year you are guaranteed entry into the following year's race. Back in the day, this was my default method of gaining entrance. I could almost always reliably end up somewhere in the top ten. But since that time an entirely new species of ultrarunners had entered the scene. These are young, wickedly fast kids that have been professionally trained, many with decorated collegiate running backgrounds and Olympic qualifying marathon times. And then there is a whole new breed of outliers, guys such as Spaniard Kilian Jornet, who could run a marathon up the side of a mountain faster than most people could cover the distance on a flat course at sea level. Over the period of a decade a new world order had emerged in ultramarathoning. Nearly every course record had been shattered in that time, and new faces were besting existing course records almost weekly. The depth of talent was deep and growing, and I was a grizzled vet straining to maintain a toehold in this sport I loved—a member of the old guard trying to keep pace.

It was demoralizing watching that young stallion bounding off into the distance, but it was reality. What little leg speed I once possessed was waning. No longer could I dependably downshift, boost the rpms, and motor up the side of a mountain the way I once could. Those days are over.

Though to my credit, I retained a certain grit. Age had slowed me, but it had also toughened me. I was no longer fast, but I was relentless. I steadily put one foot in front of the other and continued attacking the mountainside climb, step by deliberate step scaling the towering valley wall, determined not to waver.

Eventually the apex of the peak was surmounted and the course briefly leveled and turned slightly downward. The McGee aid station was at this juncture, 37.5 miles from the start. It was a remote and rugged area, accessible only by specialized off-road vehicles; thus crew couldn't get here, so no Dad. A couple of other runners were sitting around, and they looked severely baked. I wasn't sure if they were running the 50-miler or the 100K, but perhaps they'd gone out too fast. It was a mistake I'd been guilty of myself on plenty of occasions. It was easy to lock horns with another runner midway through a race and let your egos do the running. Eventually you burn yourselves out and end up walking together for the rest of the race, if you can even get that far. And I'm not sure if either of those two runners sitting there would get that far.

One of the volunteers approached and offered to refill my water bottle. I thanked him and turned to inspect the food table. All three food groups were represented: fat, sugar, and salt. It's a touch ironic that some of the fittest people on earth would suddenly crave M&Ms and salty potato chips, foods that are not normally part of the daily training diet. But put someone out on a hot and dusty trail running for countless hours and standards loosen. I once witnessed a devout vegan friend of mine stuff a fistful of beef jerky in his mouth. When survival's a factor, principles slacken.

Along with the candy, chips, and meat snacks, there was also a selection of more athletically tailored food. I opted for a handful of margarita-flavored Shot Bloks. Picture this: a miniature, shelf-stable Jell-O shot with twice the sodium and none of the alcohol and there you have a Shot Blok. They come in a variety of flavors—orange, strawberry, tropical punch—but margarita is my favorite. I popped one in my mouth and it suddenly occurred to me that whoever invented margarita-flavored Shot Bloks deserved a Nobel Prize. Brilliant. A life-altering invention (at least during the midpoint of an ultramarathon). I stashed the remainder of the packet in my pocket for later.

The volunteer brought my water bottle back. I thanked him. I also thanked him for being out here.

"No problem," he said. "Glad to be helping out."

"You a runner?" I asked.

"Used to be. I miss it."

It was a familiar refrain. I heard it all the time from ex-runners. They missed it. Missed running. Missed the magic and the misery of accelerating the human form to a place where comfort is discarded and something approaching anguish and suffering becomes the glorious, detached state of being. As I continued running down the trail, I thought about myself as an ex-runner. I couldn't form a vision of that man. I'd defined my finish line as a pine box and had no intentions of stopping until then. And what would become of me if something should happen that prevented me from running? I would cease to exist; I would extinguish and evaporate. I needed running to be complete, not like a junkie needs a drug to get high but like a seed needs water to become fully what it is.

After leaving the McGee aid station, the course took a sharp and welcoming downturn. Welcoming because I'd spent a handy amount of time conditioning my leg muscles for the downhills. As any seasoned distance runner can attest, hammering the downhills at a rapid pace is a surefire way of turning your quadriceps into hamburger meat. I know this from experience (bad experience). In the short 3 1/2 miles from the McGee aid station to the Buttermilk aid station the course descended the equivalent of the Empire State Building with the Statue of Liberty stacked on top of it. That's a lot of down in a short amount of time. And as I am maneuvering along on this stretch of trail trying to conserve my quads, whom should I come upon? Yep, my young friend. Walking.

In such moments glee can be the initial response, alpha pride, a domineering sense of hubris, oh, mighty me. But strangely I didn't harbor such sentiments. Age has given me perspective and humility. I fully recognized that it could just as easily be me who was reduced to a walker. That's how life goes. Some days you're a fisherman, some days you're a fish.

"Quads locked up?" I asked.

He seemed surprised that I slowed to walk with him.

"Yeah, they're tight. I'm trying to walk it off."

"I guess that's all you can do. At least you're still moving."

We walked together a bit longer.

"Here, have one of these." I handed him the packet of margarita Shot Bloks.

He put one in his mouth. "Oh, those are good."

"Yeah, and the sodium will help your legs."

We kept walking in the still silence of the early afternoon— two men on a trail in the mountains.

"Where you from?" I asked.

"Southern California. And Mr. Karnazes, you're the reason I'm here."

That wasn't the rejoinder I was expecting, and it threw me.

"I read your first book and that's what got me into it."

"And you're happy about that?"

"Oh yeah, totally." My statement had been meant as something of a joke, a breezy quip to release some of the tension I felt every time someone complimented me. But I don't think he took it as such.

"I want to run Western States one day, and Badwater. Yeah, man, you're super inspirational."

I've never felt comfortable accepting praise; never felt worthy. Believe me, I'm nothing special, nothing commendable. In fact, I think of myself as a bit below average. The only "talent" I may possess is doggedness. When it comes to groveling, I'm the master. Master Grovel. That's me, that's my superpower. Groveling.

"Badwater and Western States are two iconic races," I responded. "I wish you luck."

We kept walking together for a bit longer. But walking wasn't feeling so good to me; my legs needed to move. So I bid him farewell and continued motoring onward to the next aid station. He slowly disappeared in the distance behind me. Who knows? Perhaps today I'd be doing the fishing.

3

WHY WE RUN

Misery loves cutlery.

You're either born a runner, or not. Simple as that. And it isn't the act of running that constitutes this demarcation, but the desire. Running isn't necessary—not in this day and age—yet some people choose to do it. Certain individuals seek out struggle and hardship, while most look to avoid such things. Nonrunners ask, "Doesn't running hurt?" *It does if you're doing it right*, we runners answer. The mind-set of the runner is universal, our reality communal, we "get" each other. Comfort is overrated. Life is easy. Why do something difficult? Because life is easy.

Running down that hot and dusty trail at the Bishop High Sierra Ultra, gasping for air and trying to absorb the punishing impact of each footfall, it occurred to me that this could very well be another person's version of hell. Yet I loved it. It was hot, miserable, and painful—perfect, really.

I continued charging downward for the next several miles, knowing I was depleting my water bottle and overheating to the core. It just felt good, letting it out. Until it didn't. Abruptly my body felt hot. Too hot. I'd been neglectful in paying attention to my vitals and had run myself dry. I could feel a gritty layer of salt forming under the brim of my hat, my mangy, thick hair coated in a white, chalky brine. Running into the Buttermilk aid station, more than 40 miles from the start, I wasn't quite drooling on myself but was definitely sloppy, meandering haphazardly, my lips brittle and cracked, my throat parched.

One of the volunteers saw me coming and rushed over. "Can I get you anything?"

"*Water . . .*" I said faintly, "*water . . .*"

She handed me a cup and I guzzled it immediately, half the contents spilling down my chin.

"*More . . .*"

She handed me another cup and I did the same.

"*Please,*" I said, holding the cup out for another refill.

"Boy, you're really thirsty."

If you only knew . . .

She kept pouring cup after cup, but there was just no bottom, my thirst unquenchable. "We've got food, too," she added.

My eyes snapped squarely to meet hers. She must have detected a flicker of life inside, for without a word she came around behind me and put both hands on my shoulders. Then she began maneuvering me toward the food station like a forklift moving a crate. I shuffled forward like Frankenstein, stiff-legged, arms outstretched, eyes glazed.

When we arrived at the destination it was as though I'd been transmuted to a taqueria in Guadalajara. At every ultra there's always some brilliant individual who manages to get a camping stove and cast-iron skillet out to the middle of nowhere. God bless the man; misery loves cutlery.

"¡*Hola, señor!*" he boomed. "What's it gonna be?"

He was cooking quesadillas and had a complete kitchen setup going. Several options were available, and all the ingredients were displayed in separate, colorful bowls. Words would not come to me; instead I clumsily pointed to what I wanted inside: beans and cheese.

"You got it!" As he put the ingredients together and started cooking I watched, mesmerized, staring at him with a dumbfounded look as though I'd never seen a man cook before. He toasted one side of the quesadilla, then flipped it over to crisp the other. I could detect dribble running down my chin.

"Guacamole with that?"

My eyes must have widened. "Guacamole it is," he said with a wink.

On top of the guacamole he sprinkled some fresh cilantro and then squeezed lime juice all over the entire ensemble.

"Here ya go, amigo." He handed it to me on a paper plate steaming hot.

Now, I don't know what it takes to earn a Michelin star, but this guy deserved more than one. I've eaten at some fine-dining establishments, but nothing even came close. I savored every bite. It must have taken me at least forty-five seconds to finish (believe me, I exercised considerable restraint). Had no one been around I would have inhaled the thing in a single swallow the way a frog gulps a bug.

I wiped my mouth clean with the back of my hand.

"That was amazing," I said to the chef. "Now I need to sit down."

"Sure," he said. "Take your time. You running the 100K?"

"That's what I signed up for."

"You got all day. Remember the ol' saying *mind over miles.*"

I lowered myself to the dirt. "I know the saying, but I'm not sure I've got much of either left." Forty miles of running had slowly beaten me down.

Just then my young friend came dashing in. He seemed in a hurry, rushing around grabbing food and filling his hydration vest. He didn't see me sitting there and I didn't say anything. Then off he went, just like that.

This is how an ultramarathon goes, especially toward the latter stages. Energy peaks and plummets, a hooked fish seeing the gaff coming is suddenly infused with renewed fight. He'd been reduced to a walker a few miles ago, but now he was on the offensive, the body back to full capacity, the legs fresh, hope springing eternal.

Meanwhile I sat in the dirt, rubbing my temples. Yes, this is how an ultramarathon goes.

And if anybody knew about this dynamic it would be me. After all, I'd run hundreds of these things across all continents in every imaginable condition. I'd dismantled my body before only to somehow reconstitute and continue onward. I'd witnessed what the human body is capable of and had a great deal of respect for our ability to rejuvenate and resume.

I continued sitting there, massaging my quads and allowing my core temperature to cool. Another runner came in. She was muscular, her hair braided in pigtails, with several tattoos on her forearms and one on her ankle. "I'm not feeling well," she said to one of the aid station volunteers. "Is there a trash can around?"

"Sure," he said, pointing. "It's right there."

The runner walked over to it, lowered her head, and launched the contents of her guts into the receptacle. We all watched in astonishment as she retched.

She lifted her head. "Ah, that's better. I was feeling a bit queasy."

We continued staring in amazement. A volunteer finally asked, "Can I get you anything?"

"Nope, all good," she said with a nod, "boot and scoot," and off she ran.

I continued sitting there, rubbing my belly now.

Two more runners came in. One of them started foraging around the food table; the other just stood there. The one looking for the food saw me. "Oh, hey, Karno. You okay?"

"Yeah, man," I said, "just getting my sea legs back. Got a little wobbly back there."

He continued rummaging around the table, looking for morsels. "You running the 100K?"

"Yeah," I answered. "That's the plan, at least. You guys?"

"We're running the 50. This is Kevin's first."

Kevin continued just standing there with this perplexed, shell-shocked look on his face. I wanted to snap my fingers to bring him out of it.

"How's everything going, Kevin?" I asked.

He began muttering something unintelligible, seemingly getting hung up on the first syllable. "*Aye . . . aye, aye . . .*"

The forager finished packing his vest with food. "Okay Kev, let's roll, buddy." He took hold of Kevin's arm and spun him around. "We're going that way."

"*Aye . . . aye . . . aye . . .*"

Away they went.

I continued sitting there rubbing my temples, counterclockwise now. Random thoughts entered my mind. I thought about my family, my kids. I thought about an old friend at school.

I thought about the pricker sticking out of my sock. I thought about how much I loved this damned sport, how thoroughly I dug every second of it, everything, in its totality. Even sitting here on my rump at this aid station watching the action unfold. It was good theater, I tell ya. The characters came and went, each playing their individual part in a unique act. Entertaining stuff.

Though when I came around to thinking about it, I hadn't come here today to be a spectator. No, I'd come here today to be part of the show, to play my own role in this unfolding saga. Enough with sitting in the audience; it was time to get back on stage.

The good news was that I managed to rise to my feet without having to first take the intermediary step of rolling onto my belly and then crawling to an upright stance, one knee at a time. It was much more dignifying not having to do that. And don't you know every eyeball in the aid station was tracking me to see which way it'd go. Those vultures! (Though I would have been doing the exact same thing myself had I been in their shoes.) But I got to my feet with self-respect intact.

The first few steps out of the aid station were a little tender, but eventually the needles stopped pricking. With 40 miles of running on the legs there's bound to be a few niggles. At least all my toenails were still attached. The course continued its descent, not too steep, not too gradual, the footing good, if not overly dusty at times. Several miles passed, and the lower the trail proceeded, the hotter it got. About three degrees Fahrenheit for every thousand feet of elevation change, I believe. They call it the adiabatic lapse rate, and it's precisely why you've got to be cautious at places such as the Grand Canyon. The adiabatic lapse can sneak up on you. Hike down to the valley floor a few thousand feet below the rim and suddenly it's fifteen or twenty degrees hotter, and now you've got to get back up. Plenty of heatstroke victims are medevaced to safety every year due to this phenomenon.

A few more miles passed, and I didn't see another runner. We appeared to have each established our own cadence and were being carried downstream like flotsam in a current, occasionally bunching up and occasionally coursing forward, each drifting independently toward the valley floor. Gradually I got swept up in swifter-flowing waters, the energy ebbing back into my system, calories from the recent fiesta converting to aerobic output. My stride sharpened and my footing steadied. I seemed back.

Off in the distance the Highway 168 aid station came into sight, a small red speck in the foreground set against the immense Owens Valley with the austere White Mountains jutting skyward on the opposite side of the basin, their colossal granite crags visible like cavernous tectonic plates. The air was still, perfectly still, ripples of heat rising from the valley's floor. There was no sound. No sound except for my breathing and my murmuring affirmation, *"You can do this . . . you can do this . . . you . . . can . . . do . . . this."*

Slowly as I preceded toward it that small red speck in the distance morphed into a larger dot, which ultimately became the red canopy covering the aid station. The Highway 168 aid station is readily accessible to crew, and the first person I should see as I came running in is Dad. He had a foldable beach chair set up next to our cooler in the shade of the tent.

"Hiya, Pops," I said.

"Hello, son. Keepin' your nose clean?"

"I'm trying to, though I've had my moments."

"That's to be expected after 46 miles of running. Would you like to sit down?"

I gave him a disapproving look.

"Oh yeah, *beware of the chair*," he reminded himself. Sitting during an ultra can be a one-way descent. Better to avoid the prospects altogether unless you're completely destroyed (actually, *especially* if you're completely destroyed).

"Well then, can I get you anything?"

"Do you have that pepperoncini juice handy?"

Pickle juice is all the rage with long-distance runners. It helps to alleviate muscle cramps, they say. But just like running, you're either a pickle person or you're not. Personally, I can't stand the stuff. Let me be fair to our wrinkly little green friends: I don't mind the occasional pickle, but drinking, and inevitably burping, pickle juice is, frankly, disgusting. I find pepperoncini juice much smoother. It's just as efficacious—I would argue more so—and it doesn't leave the aftertaste of fermented cucumbers in the back of your throat.

Dad reached into the cooler and handed me the bottle of pepperoncinis. I unscrewed the lid and took a swig. "Ah, mother's milk."

He gave me a ghastly look.

There were about a dozen people hanging out around the aid station. Most were volunteers or crew, but there were a few other runners. One of them was my young friend. He was sitting in a chair at the far side of the aid station eating some food. He didn't appear as damaged as I did, though looks can be deceiving. It's what's inside that matters; every runner knows this.

I turned to Dad. "That kid's tough."

"He's young," Dad offered. "He's been coming in ahead of you to most of the stops."

"That's the future, Pops. He's way faster than me, and he knows how to run. The next generation's changing everything; it's becoming a different sport now. But enough of that. How are you?"

"Me? Just fine," he said. "In fact, if I were any better I'd be twins."

I guffawed. "I dunno where you come up with these things. Just stay hydrated. Both of you."

I took one more pull from the pepperoncini juice bottle and handed it back to him. "Thanks for everything, Pops, it's really nice having you here with me. Just look at that Owens Valley."

We both turned to marvel at the sweeping expanse before us, briefly transfixed by the grandeur of the surroundings. So many years had transpired since we first came here, so many memories accrued, mostly good, though some devastatingly painful. I guess that's what happens when a life is lived long. You accumulate a trove of lasting memories—a beautiful collection of reminiscences that fills your spirit with joy and warmth—but you also endure loss and sorrow, and it's during these reflective moments in vast expanses when you think most of both.

"Say, Dad, is it hard for you seeing me like this, all beat up and struggling?"

I'd never asked him this question before; our relationship rarely veered into such personal territory.

I noticed him pause for a second, something uncharacteristic. "Yes, it is. No father likes seeing his child in pain. It doesn't matter how old you are, you'll always be my son."

His words hung heavily in the air; having two children of my own I knew these feelings. A child's pain is a parent's pain, magnified.

I patted him on the shoulder. "I'll see you at the finish line, Pops."

I set back out, nodding to my young friend as I ran off. He nodded back, still chewing. He looked anxious; I didn't anticipate he'd be spending much more time in that chair. But we each had our own race to run, and much distance still left to travel.

The finish of the 100K was some 16 tough miles ahead, and given the difficulty of the terrain, and the heat, and the remoteness of the remaining course, along with the cumulative bodily damage thus inflicted, covering this stretch could easily take five or six hours, maybe more. Yet I knew, with all certainty, that Dad would drive directly to the finish. It would only take him perhaps twenty minutes to arrive, but that is what he would dutifully do. He would drive to the finish and wait patiently for me. Even if it took me nine or ten hours and went well into the night, he would be there, waiting. Semper fidelis. Always faithful.

Almost immediately after leaving the Highway 168 aid station a new sort of fatigue emerged. My legs felt heavier and more lethargic than earlier, as though they now mysteriously bore the burden of ankle weights. Lifting my feet high enough to clear the ground seemed overly draining. For the first time today, I felt drowsy, genuinely tired, and past the point when caffeine would reboot the system. It was all catching up with me, and quickly. Lead leg is not something typically recoverable; it's the kiss of death for runners, a wholly incurable affliction. And the onset was troubling, so quickly had I gone from slightly wounded to seemingly terminal.

Just get to the next aid station, I told myself. *Hold yourself together for a short couple of miles. You've been here before; just get to the next aid station.* There was nothing glamorous about the way I got the job done. The trail was still relatively groomed and downhill, yet I kicked up a plume of dust like a lumbering mule train, any remaining spring in my stride sprung, my hoofs incapable of clearing earth. It was the rapid onset of my deterioration that was most troubling, and I damn well knew the cause. I wasn't trained for this—not this distance. It was too much to be taking on so early in my ramp up.

I kept grinding along. There are two ways to get to the finish of an ultra. The first is to put your head down and grunt your way through it. I don't know the other way.

After considerable toil, I arrived.

The Tungsten City aid station was strategically situated at mile 48.5 along the course. Despite the name, there was no "city" in Tungsten, just a small tent and a few volunteers hanging around, greeting runners. I must have been quite a sight to behold as all of them snapped to attention when they spotted me stumbling in. On this occasion I gladly accepted the chair offered me, no second thoughts. I plopped down, moaning audibly, and sat slouched over, my mental acuity flickering sporadically like an old neon sign.

The volunteers brought me all manner of replenishments— liquid and solid—but I wasn't in the mood. Nothing seemed right. I just wanted to sit there and think for a while; actually, *not* think was more like it. I was rudderless, no sense of direction, uncertain as to what my next move should be, not enough energy remaining to control the central governor (i.e., my noggin).

"How you doin'?" a voice reverberated in the empty space between my ears. It was a man's voice—that I could detect—but where he stood was unclear. "You with me, partner?" the voice asked again, a gentle hand on my shoulder.

I seemed unable to elicit a verbal response, so I just shook my head. Whether that was a nod of affirmation or denial was anybody's guess. I'm not sure I knew myself.

"Here, have some of this," the voice said. I saw a cup being raised to my lips and I gulped intuitively. It was Mountain Dew Code Red (the "Code Red" signifying an overabundance of sugar and caffeine). It took a moment to process through my system, but eventually the tonic worked its magic, and feelings slowly returned to my extremities. I had another gulp, the fog lifting slightly more.

The voice continued, "Hey, so it looks like you're signed up for the 100K, but the 50-miler's still an option."

"Yeah, so is a DNF," I mumbled.

I couldn't tell if he was serious. I'd never heard of such a thing, but he made it sound like it was, well, a thing (like, duh man, how could I not know this?).

"Really?" I asked.

"Really," he responded. "Others have opted for it. You can trade down from the 100K and still get credit for the 50-miler."

There was a protracted moment of quiet. I couldn't answer. I was unable to answer. I couldn't comprehend. He went on.

"If you make a right turn here it's a mile and a half to the finish of the 50-miler. Done. If you make a left turn it's a long path into desolation."

I looked at him. Then looked down at the dirt. Then looked back up at him, more like through him.

"So whaddaya gonna do?" he inquired.

"Weelll," I said slowly, "I really didn't sleep much last night, and I'm really in no shape to be out here, and I'm really not feeling that well, and my dad's already waiting for me at the finish, and . . . and . . . and . . ."

I went silent and there was only a hollow, vacant quiet.

Finally he shattered the calm. "So whaddaya gonna do? 50 miles or 100K?"

"Well . . ." I said.

Another long pause.

"Well?" he questioned.

"Well . . ." I started to answer.

4

FOLLOW THE PATH

You needn't fear the darkness if
you have the light inside.

My friend Nick Moore (aka: Preacher) has a saying about moments of profound reckoning, when exhaustion is so acute one must summon a higher authority for guidance. He refers to such instances as "Come to Jesus" moments.

Sitting in that chair at the Tungsten City aid station, teetering on the precipice of incoherence, I had such a moment. The conversation was portentous and went something like this:

You must follow the path, the voice instructed me.

"But Father, the path is filled with darkness."

You needn't fear the darkness if you have the light inside.

"But Father, how do I know if I have the light inside?"

Follow the path.

I continued sitting there, still not ready to move.

"Well," the volunteer said once again, "whaddaya gonna do? Fifty miles or 100K?"

"Well . . ." I started to answer. "Well, forget about all that stuff I said earlier about being tired and not in shape."

He looked surprised. "Those were just excuses," I said, "and excuses never got me nowhere."

"So you're gonna run the 100K?"

"Hell yes I'm gonna run the 100K!"

Now he looked either excited or amused. Actually, both. "Okay, soldier," he said in a jocular inflection, "I'll top off your bottle and send you on your way."

He enthusiastically refilled my water bottle and thrust it into my hand.

"Thanks, man," I said. "And one over the head?"

"Yep." I pulled off my cap and tilted my head backward. He grabbed a cup of water and poured it slowly over my forehead as I began rubbing the cool liquid around my face and neck, like an abundantly generous dousing of Old Spice.

"Ahhh . . ." I said, shaking the excess like a wet dog. "That's nice."

"Nother round?" he asked.

"Sure."

We repeated the exercise and I allowed the second cup to run down my shirt.

I continued sitting there, not wanting to get up, which made forcing myself to do so all the more gratifying. Oftentimes it isn't talent that gets us someplace, it's bullheadedness.

"Good luck, man," he said as I began heading off into an uncertain future, and he said it like he meant it, like I would need it, for the high country of the Sierra Nevada isn't known for lavishing luck upon passersby, more like just the opposite. Plenty of settlers met their doom along this sprawling hinterland. Let's hope I have the light inside.

The terrain quickly changed from a soft, gradual downhill to an abrupt, rocky uphill. Almost immediately I was reduced to a walker, and then to a slow walker, which, not so disagreeably, afforded me a chance to look around. I didn't see much other than barren, undulating desert knolls in every direction. While there'd been few people on the 50-mile route, there was absolutely nobody on the 100K route, not the slightest hint of a lingering dust cloud to indicate someone had recently passed, and only a random footprint in the dirt indicating anyone had ever followed this path in the distant past. I felt very alone. Rightly so, because I very much was.

I took a big pull from my water bottle. The liquid was already warming. Why only one water bottle, and an uninsulated one at that, speaks to my imprudence. When packing for an ultra it's often easy to fall into the trap of underestimating the severity of your undertaking. Sorting your bags in the comfort of your home, where you're operating in a pleasant state of homeostasis—not too hot, not too cold, amply fed—forecasting your state of being midway through the race doesn't come intuitively and it's easy to apply your current preserved and fresh state of bodily contentment to your preparations. This, of course, presents a false sense of security that can lead to gross oversights.

Something in me seemed prone to such unpreparedness. Perhaps it was my quest to keep things novel and unsure in a world that seemed so thoroughly overregulated and safe. Just throw yourself into it completely ad hoc and see what happens. Fun! Fun! Or perhaps it was just laziness; I couldn't be bothered with details and just wanted to run. Either way, here I was with one measly uninsulated water bottle when the conditions called for at least two big insulated ones. Better yet, a large hydration pack.

I took another draw from my pitiful little water bottle. It

would have been prudent to drizzle a little over my head to prolong the moisture from the aid station drenching, but I didn't
have enough to spare. Conservation was necessary. That water
bottle and I would have to become intimate friends over the remaining 13 miles.

Progressing farther, the 100K course was absolutely unrelenting. Whereas earlier there'd been long, drawn-out climbs, now
they came in rushed, steep pitches exacerbated by acute, severe
declines. It was impossible to run either. The climbs were too
vertical and the descents too sharp. It was slow going, slow and
arduous under a remorselessly blistering sun. *Only a half marathon*, I kept telling myself, *just 13 miles*. But that was self-trickery
I couldn't lure myself into. I knew better. Sure, with 49 miles
under my belt I was proportionally a good deal done with the
race. But in absolute terms, I still had a brutal and demanding
half marathon to go. One has plenty of time to think during an
ultramarathon and deception finds few hiding places. You can
only fool yourself for so long before stone-cold reality ultimately
prevails. And that it always does. Nothing sharpens the focus
like self-preservation.

It took a long period of disciplined trudging to arrive at the
Sage Summit aid station, 52 miles from the start. It was a reliev-
ing sight, as what little water remained in my bottle was warm
and stale. The Sage Summit aid station was located in an isolated and desolate region, and the three aid station attendees had
hauled everything in on ATVs. They were drinking beers when
I pulled in but sprang to attention when they saw me coming.
"Howdy there," one of the party greeted me.

I tried to mount a response but could only croak something
incomprehensible. "Here," he said, handing me a cup of water,
"have some of this."

I gulped down the water and thanked him, then looked around at their setup. It was clear that there hadn't been much traffic through this aid station as the food was still plentiful and neatly arranged on the table, and the three attendees seemed all too eager for a change of pace. "Can I get you anything?" another one asked. He was an affable lad with glowing, rosy cheeks and a big, bushy beard. All of them wore floral Tommy Bahama shirts and straw hats.

"Got any ice?" I finally managed.

"You betcha. We got a coupla YETIs full."

That was welcome news. Had he said they had a couple of *coolers* full I would have worried. In this heat, ice melts quickly, even in coolers. But YETI coolers come from different stock. These babies were created for conditions precisely like these.

"You guys are like a gift from heaven," I said. "You've even got a long beard."

They found my comment humorous (clearly they'd been out here a long time).

"Look," I said, "if I'm gonna get through this I'll need to load up on ice."

"Sure thing, partner," he replied, "we haven't had many visitors today." I'd guessed as much, but didn't say anything.

He opened one of the coolers. "C'mon over."

Inside was a metal scoop and he filled my cap, which I promptly slapped on my head. Then I had him shovel a scoop into each of my front pockets. "I'll never have children again, but at least I'll stay cool."

They found that comment amusing. Clearly it'd been a long day. He then filled my water bottle and screwed the cap back on.

"Okay," the bearded man continued, "you've got about 2 1/2 miles to the turnaround. They won't be easy. You'll drop about twelve hundred feet in that time (translation: *It's damn steep*). When you get to the turnaround point you need to look for a little stick in the ground. It's kinda off to the side of the trail, I'm told, so be alert. When you find the stick, go to it and look around. Down at the base there should be a sheet of smiley-face stickers. Unpeel one of the stickers and put it on your race number. Then reverse course and head back."

I looked at him with my mouth agape (translation: *You're shitting me, right?*).

He went on matter-of-factly. "There're a bunch of different colors, just choose whichever one you like."

I stood there in disbelief. "Let me get this straight," I said, "I'm to run off into that godforsaken wasteland in hopes of finding a little stick with a packet of smiley-face stickers next to it?"

"Yep," he replied, "that's about right. Put one on your bib so we know you made it to the turnaround, then come back. Choose whichever color you like."

I kept waiting for them to burst out laughing. But no, they just stood there looking at me, like, *Well, you gonna go?*

"So these stickers," I said, "are they like the ones you get in grade school for a good homework assignment?"

"I think so," the bearded man answered. "One of the volunteers is a teacher or sumpin'."

I kept waiting for the laughter, but it never came.

So I started reluctantly moving forward, exiting the aid station. "Choose any color I like, right?"

"Yep. And remember to watch for that stick. You don't want to run past it."

"Riiiight," I said slowly. "Guess I wouldn't get a very good grade if I did that."

"No, you wouldn't," he said, "and we want you to get an A

plus, so don't run past the stick." Of course, running past the stick meant more than a failing grade, it meant possibly dying an excruciating death of thirst and starvation lost in the desert and having your eyeballs pecked out by vultures.

I couldn't believe this conversation. Moreover, I couldn't believe that after 52 miles I was about to run off into the desert hunting for a little packet of smiley-face stickers. But that is precisely what I was doing as I set out.

One thing he was dead serious about is that twelve-hundred-foot descent. It was no joke. Fortunately, there were switchbacks cut into the side of the grade that helped lessen the severity of the overall pitch. However, it faced west, which meant the afternoon sun beamed directly onto the forehead of the slope. And there wasn't a tree in sight, not so much as a small shrub. There was only a harsh Mars-like dusty red planet as far as the eyes could see.

The ice in my pockets eventually melted, as did the scoopful under my hat. It was alarming how quickly my clothes lost their wetness, how transient the miracle of evaporative cooling had proven. By the time I reached the bottom of the descent I'd completely dried. There were no more tricks left up my sleeve; it was now man against the elements.

The trail flattened at the base of the downslope and proceeded across an everlasting open plain, soaring mountain peaks lining the distant backdrop. I kept following the path onward deeper into the desert. As I did, it became less of a path and more like a wild goat trail. This is not a frequently traveled corridor, not at least by two-legged animals. I kept running across this open expanse trying to discern the correct route among the many dissecting offshoots, all the while looking for a little stick in the ground. Finally I gave up.

I must have run right past it. Or did I? Suddenly I wasn't so sure. Is it still ahead of me, or is it behind me? I stopped and looked back. Then I turned and looked forward. It all looked exactly the same. Hmm . . . should I turn around or should I keep going? I could feel the vultures circling in.

It was maddening not knowing what to do. And it was insufferably hot and motionless. I looked around some more: nothing. I finally concluded that I should keep going. I wouldn't turn around until I found that damn stick and plastered one of those smiley-face stickers on my race bib. If I'd run past it, so be it. I'd just keep going until they sent a rescue party after me (wishful thinking that would happen anytime soon). Maybe the vultures would spare me, finding my fatless body unappetizing, and fellow humans would one day come upon my lifeless remains out in the wild, mummified and rigid, though perfectly preserved by the arid desert air like King Tut.

Onward I went. Onward farther. Nothing. Nothing at all. I wanted to scream. And why not? "AAHHH!!" I bellowed. It felt good. So I did it again. "AAHHH!!" I ran in a crazed drunken bloodlust like some inebriated feral beast, hunting for something I wasn't convinced existed, not at least on this particular offshoot of the trail I was currently traversing. Or perhaps it did, perhaps just ahead I'd find that stick. I kept craning my neck searching, scanning the landscape looking for anything that resembled a thin, cylindrical shaft. Nothing. Faster I ran, still faster. I was possessed; I'd come all this way, invested far too much of myself to be denied. I needed to find that stick! *I should probably watch the ground*, I thought, *there are rocks and roots all around, this is how people get hurt, I should probably pay attention, I should probably slow down.* But it was too late . . .

The tripping and falling part wasn't the problem; it was the landing that caused the issues. Cartwheeling through the air, I came crashing to the earth like Icarus, abruptly and unexpectedly, skidding to a stop on my belly, my chin digging a little divot in the dirt, left arm pinned beneath me, right arm skewed off to the side. Lying lifeless, heaped headfirst in the soil with the wits knocked out of me, I simply watched as small whirls of dust spiraled out from my nostrils with each labored exhalation.

Until I accidentally inhaled too forcefully and started choking on the dust, at which point I used the left arm pinned beneath me to flop myself over in a manner a short-order cook uses a spatula to flip a pancake. Now I'm lying splayed ass-to-the-dirt in the trail—one leg tweaked improbably beneath me—staring up at the afternoon sky, seeing stars where there are none, wondering what the universe would be like without gravity, or at least a little less of it. Upon tripping, would I have just kept flying? Or, like a scene from *The Matrix*, would I hang in a state of suspended animation just long enough that I could get my feet back under me and continue running? Funny what goes through the mind when out for a ten- or twelve-hour run.

Eventually I came back to practical issues, such as *Is anything broken?* Or, *Will anyone find me?* And, mostly, *What the hell am I doing here?*

I reached down with my hand and felt my stomach, moving over to my ribs. They felt sore, but more bruised than broken. Would anybody locate me out here? Doubtful. So rather than risk waiting, I thought it prudent to move. Best to drag myself off the ground, piece things back together, and resume forward progress.

So much for best-laid plans. I continued just lying there. It simply felt too good not moving (more like moving just felt too awful). Lying there was more or less an act of avoidance, of putting off what had to be done.

I continued to ponder the fundamental question of what I was doing here. Of any question to pose to oneself, this one is surely the most disquieting. Whenever self-doubt creeps into the equation the stakes get critically elevated. Let's face it: I was no good at this anymore, these long races had become such struggles. My younger self could waltz through a 100K, but now such contests invariably digressed into Homeric slugfests. Maybe I didn't belong out here anymore; maybe it was time to quit. I continued staring at the sky, contemplating my reason for being.

This was one of those bigger life questions: What is one's purpose? Had an earlier calling run its course? Was it time to let go rather than cling to the memories of what once was? For me, this was a deeply unsettling inquisition. My entire existence was built around the sport, activity, and lifestyle of running insane distances. Trite as it may be to some, it's all I had.

But this was not a lighthearted subject and something best left repressed for the time being. Now, here in the throes of battle, what I needed most was the unwavering tenacity to remain entirely in the game. A wandering mind can be a dangerous thing during an ultra; my attention must stay confined to the immediate task at hand: finding that little stick in this ever-expanding desert. It was with this singularity of focus that I arose from the earthen dust and resumed searching.

Largely in vain. All I saw was a titanic landscape of sameness. This was the desert, after all. How on earth was I supposed to locate something so ubiquitous as a little stick in this barren sea of rocks, roots, and high desert flora?

And then I spotted something out of the corner of my eye. I turned to look. Could it be? It was off to the side of the trail by some distance, but it looked stick-like. Pensively I jogged over toward it, not wanting to get my hopes up. But miracle of miracles, it was "da stick." I screamed, "Eureka!" like I'd struck the mother lode. Then came the matter of locating that infamous packet of stickers. But sure enough, they were right there. *Well, whaddya know*, I said to myself.

I peeled off one of the smiley faces and stuck it to my race number. Then I stuck on a few more for posterity. The packet was barely touched, and I was relatively certain there wouldn't be hundreds of runners venturing out to this spot today (or ever, for that matter).

With smiley stickers attached, I faced the vexing task of heading back to the Sage Summit aid station, which meant scaling that same twelve-hundred-foot monster I'd descended on the way out to this turnaround point. I started running toward the beast on the flatland approach trail, staring down the creature as if tiny me could somehow intimidate this hulking mound of rock, with its diagonal switchbacks lacerating its exposed face like gash wounds from a wolverine. It looked angry displayed before me, like a fierce opponent entering the opposite side of the octagon. People say running is a battle against self. I would say this is true to a degree. In track and cross-country, let's be honest, it's you against the competition. Perhaps in the marathon, a long contest where many of the variables are held constant, it's up to you to persist through mile after mile of racecourse, trying to suck energy from the cheering crowd, but knowing ultimately that the battle lies within. But here, during a protracted and exposed ultramarathon, where man and woman are pitted not just against great distances but also against the savage wild in all her spirited moodiness, it's largely a battle against the elements and against the land, you versus all the extremes nature can hurl your way.

"ARR!" I screamed at the mammoth upslope that I was about to engage in combat. Squaring off with a vertical mountainside is overwhelming and horribly intimidating, I'll be candid. You can see the top, but that's part of the problem: it's impossibly high up in the sky. It was a formidable adversary I was up against, and this would be nothing short of warfare. At its best, trail running is something of a dance with nature, but at times like these it becomes more like a visceral fistfight. To emerge victorious I needed to conquer this enormous challenger, and that could not be accomplished with a single powerful blow. No, getting to the top would require a calculated, systematic dismantling, a series of repetitive body blows that slowly chipped away at the incline. I put my chin down, snarled, and started the long, hot ascent, not as a runner but as a disciple of perpetual forward momentum. *Keep moving forward*, I told myself, *and hope the light is inside.*

5

CAN'T STOP; WON'T STOP

If running's a drug that threatens
my life, let it have me.

Inside every tame man is a wild beast yearning to get out. We'd become domesticated in our time, with our institutions and paved-over earth, our iPhones and internet, our fitness trackers and group spin classes. Gone was the uncultivated human experience. There was little solitude left in this world, few chances to truly escape humanity's trappings and feel the realness of what we are.

Running provided that gateway, and I much preferred it to a contented modern existence. Out here I was more alive, I could feel the movement of my body and the beating of my heart, I was in touch with my breathing, and my senses seemed more acute and in tune with earth's rhythms. Running great distances was a means of purging the modernity from my conscious, of rinsing the outer man from my skin and letting the inner animal reveal itself. It was at times like these when I could see most clearly, when everything within me came together and life felt very true. It wasn't every day, but today, running on this trail, I was genuinely alive.

Ascending this grandiose uprising I could hardly maintain upward progress at points. The pitch was so steep I doubled over with hands on knees trying to catch my breath, my body feverish though not producing any sweat. Crystalline salt grime coated my legs, the sleeves of my jersey felt coarse and scratchy against my arms. I could feel the brim of my hat abrading my forehead, any final particle of moisture entirely desiccated by the intense solar radiation. The mountainous face was formed like a massive satellite dish, and the lowering sun concentrated all its intensity directly onto the center of this immense earthen pockmark, searing my skin even through the protective layer of my clothing, as though I were a bug being zapped by a gigantic magnifying glass.

This probably isn't healthy, I thought. *This is probably killing me.* But it was also giving me life. Every step killed me a little, and every step gave me a little life. Onward I labored, gasping for air, my breathing strained and shallow. Perhaps I was cutting life short. Perhaps so. It didn't matter, really. I didn't run to live longer; I ran to live fuller. It may be killing me, but it was on my terms, I was doing what I loved, there is no finer meaning to be found in life.

The trail just kept going up and up, disappearing forever into the sky. The air was oppressively still, void of any headwind, tailwind, or crosswind—no movement whatsoever. I slogged along, the dust from my feet wafting languidly the way hot gas bubbles up from a lava field. There was no hiding from the elements, no possibility of deflecting the rays of energy or shielding myself from the sun's electromagnetic emissions. I was exposed and vulnerable, a settler that'd strayed from the wagon train and gotten himself into trouble.

These are the moments I live for. Perhaps no other sport holds a mirror to you the way running does. Running exposes your inner self with unvarnished brutality. How do you respond when the tide turns against you? What do you do when the going gets tough? It's been said that without war we do not know if we are cowards or heroes. The runner knows this truth, for the runner has waged war. "AROO!" I howled in the air the way a Spartan warrior would when steadying himself in combat. "AROO! AROO! AROO!" I bellowed.

There are degrees of deterioration that take place during an ultramarathon, and I was approaching the final phase. I didn't care. I was all in, completely committed to fighting it out until a victor emerged. I would prevail or otherwise, reckless as it may be. As Patton notoriously said of war, "God help me, I do love it so."

My footsteps became unsystematic; periodic progress was made, but it was followed by episodic staggering to and fro. My gait was that of a drunken seaman in rough waters—I couldn't hold an upright posture for more than an instant or two before listing to starboard. "Yo ho, yo ho, a pirate's life for me," I mumbled under my breath. The talking to myself helped keep me lucid; it's when the voice goes silent you start to worry.

In time Goliath was slain. But it was warfare, both of us bloodied and bashed. It was hard to say how long the fight had lasted—time warped at this stage, bent around itself and melted. I didn't look at my watch for fear of what the numbers would reveal. *Just keep your eyes on the horizon,* I told myself, *maintain proprioception, keep the compass level, steady as she goes.* I could see the Sage Summit aid station up ahead, but it was through ripples of heat rising from the ground, an amorphous mirage seemingly moving farther and farther away with every advancing footstep. Then, suddenly, the bearded one spoke again. "Whoa," he said, "what happened to you?"

I gazed at him emptily. "Five miles of running across hell happened to me," I slurred, as though just coming out of a deep trance. How I was now standing at the Sage Summit aid station was beyond me.

I tried to elaborate, but I was deteriorating quickly—"Eeetz hot ouwt therz." Incoherent gibberish spewed forward, the way a dying computer spits out lines of gobbledygook before permanently crashing.

I cratered into a chair, and it engulfed me the way a sea anemone ensnarls a hapless hermit crab, tentacle by wispy tentacle, pulling me deeper and deeper into the paralyzing nematocysts. I didn't think I'd ever get up from that chair; I would be stuck to it and slowly digested.

One of the volunteers spoke, but his words reverberated and echoed as though we were in a cave:

"Hello in there . . . *in there . . . in there . . .*"

"Is there anybody home? . . . *home . . . home . . .*"

I couldn't mount a response. Another one of them approached with a cup of coconut water. He held it to my mouth, but I just stared at it blankly. After a few moments of my witless gazing he finally put his index finger on my forehead and pushed my head back until it tilted slightly upward. Then he put a thumb on my chin and pulled down my jaw to open my mouth. He raised the cup to my lips and poured a little in. I gurgled.

"Good," he said, "that's good." He spoke to me as if I were an infant.

He poured a bit more in, and this time I gulped audibly.

"That's right," he said. "Just take little sips. That's a good boy."

I wanted to smack him, but I lacked the energy (not to mention he was much bigger than me). Though I did appreciate him speaking at a level I could comprehend. He kept feeding me the liquid, and eventually most of the cup was drained. I kept sitting there, and they kept looking at me to see what was going to happen.

"Well . . ." I said to them. "Well . . . I gotta go!"

And just like that I popped to my feet, like Uma Thurman in *Pulp Fiction* getting a syringe of adrenaline stuck in her heart. They looked at me in amazement (hell, I was just as amazed). But that's what happens during an ultramarathon. You have periods when you feel you could run forever, when you're invincible, followed by periods when you can barely move. At the later stages of a race these episodes become more intense and more concentrated. Eventually you simply give yourself up to it; you lose yourself in the madness of this illogical sport and hope for the best.

"Looks like you got your smiley faces," the guy standing behind the table said, looking at my race number, the top buttons of his flowery shirt hanging open. "Can we get you anything else before you go?" he asked.

"Ah . . . maybe some more coconut water?"

"Whaddabout a piña colada?"

"A piña colada?" I looked around at the dusty, dry desert. "Suurrre, I'll have a piña colada," I said cynically.

He reached into the cooler. "Here, we just made up a fresh batch." And just like that, he handed me a friggin' piña colada. Right there in the middle of the desert, a friggin' piña colada. Not only was it icy cold, the damn thing even had a wedge of pineapple garnishing the brim and a maraschino cherry on top!

I looked at him as though I were watching a magic show. "How—on earth—did you do that?"

"We got a blender," he said, "and we got music and lights for tonight."

"How?" I pondered.

"We got solar panels and a generator."

"You do?" I said. "Out here?" I paused to think. "Maybe I don't need to go after all." They laughed at me (but I wasn't kidding).

I had a sip of the piña colada. Then another. With fewer than 5 percent body fat and having run all day, I could tell pretty quickly it was a *real* piña colada.

"Boy, that's a . . . a serious drink!"

He pulled out a bottle of Captain Morgan rum and said with a wink, "Aye, aye, matey."

It was all too perfect. I kept waiting for the director of a TV commercial to pop on set and yell, "CUT! We nailed it, people; that's a wrap." This scene was somehow better than real life could possibly be.

It didn't take long for me to polish off that piña colada. "That was unfriggin' believable, you scallywags!"

I started to turn but rocked sideways and nearly tipped over. Two of them rushed to either side to help steady me as I prepared to commence running. "Is this even legal?" I asked.

They thought that was funny, too. Those poor, poor men. Clearly it'd been a long day, and it was sure to be a long night as well. Good men, these. Very good men.

"Gents, it's been a real pleasure." And off I ran.

The transition to running wasn't so terrible. My entire lower body felt anesthetized. My head didn't feel so bad, either.

I was in a better place now; some inexplicable circuit breaker had tripped and I was renewed. The piña colada and the levity of the aid station had certainly helped, but I could feel something more powerful. I could feel an energy force coming back to me.

But as I've explained, these instances can be fleeting. Just as quickly as the lights flicker on, the lights can flicker off. Thus I ran hard and fast, I poured on the gas (and I'd just swallowed some rocket fuel to supercharge the engine). The Sage Summit aid station was approximately 5 miles from the finish, less than a 10K, I told myself. Let 'er rip.

And that I did, sprinting at near top speed, tearing down the pathway without reservation. The afternoon sun was now casting shadows behind some of the larger peaks, and for the first time today I could feel temperatures cooling, the dimming rays no longer piercing the way they had the previous 57 miles. Faster I ran.

The final aid station was at the 60-mile mark. I didn't even stop. Hardly slowed. I thanked the volunteers as I ran past and just kept going, the magnetism of the finish line drawing me in. In those final few miles of running, I didn't fatigue, didn't exhaust, didn't falter; if anything, I grew stronger. The human body is far from understood. If we can just suspend our skepticism and believe in miracles, sometimes they come true. Not long ago I was facedown in the dirt, unable to move; now I was running at a hardy clip, entirely resuscitated. Yes, miracles do happen.

With about a mile left, the course broadened from a narrow, rocky trail to a wider, gravel footpath. I could see in the distance a swath of humanity where the finish line was located, and at my current pace I'd be there in about seven and a half minutes. Onward I bounded, impervious to the past 61 miles of running, fresh as a man just entering the arena, the light glowing brightly inside.

As I came into the finish area there were a handful of people milling about. Some appeared to be stragglers who'd finished one of the shorter race distances earlier in the day, and others looked to be nearby campers out for a late afternoon stroll. Some looked utterly confused, as if they had no idea a race was taking place. A few clapped and offered congratulatory words as I ran past and I nodded my appreciation. It was Saturday evening, and in some other sports universe—in a big, ostentatious stadium—there was likely some insanely paid athlete being cheered or jeered by throngs of rabid fans, the gamesman's every move scrutinized and dissected by broadcasters and commentators, and then replayed on giant JumboTron screens to rebroadcast the critical moment and reignite the crowd's fervor. This was not that place. This stadium was a few people sitting around on beach chairs having a beer as I ran across the finish line.

One of them came walking over to me.

"Congratulations," he said. "You're first in your age group and fourth overall."

He handed me a medal. "There's a barbecue over there and a cooler with some beers. Help yourself."

I thanked him, and that was that.

"Oh," he added, "your dad's welcome, too. He's a good guy."

And sure enough, there was Dad. He had loyally waited for me at the finish line, a faithful companion standing nearby in anticipation of my arrival.

"Hiya, Pops," I said to him.

"Ultramarathon Man!" (I gave him that look, like, Dad, *pleeease*, but I knew it was no use.)

We hugged. We always hugged nowadays. It wasn't always that way—not when I was younger—but that changed following an unfortunate event. Now we always hugged.

Together, arm in arm, we walked over to the barbecue area. I looked at my medal and it had "20 miles, 50K, 50 miles, 100K" printed across the bottom. It didn't matter which of the races you did: everyone who crossed the finish line got the same medal. I loved that. I'm sure someone had struggled just as hard as I had to finish the 20-miler. Why did I deserve anything better? Running is the most democratic of sports, and ultrarunning ever the more so.

The folks at the barbecue welcomed us and we exchanged pleasantries. Whatever they were grilling, it certainly smelled like they knew what they were doing. I'm sure it was delicious, but I didn't have much of an appetite, not immediately after that 6-mile dash to the finish. "Dad, do you mind if we head back to the hotel?"

"Sure," he said. "You're not hungry?"

"I'd like to take an ice bath." I thought for a second. "Actually, I wouldn't like to take an ice bath, but I need to take an ice bath."

The grillers all agreed that an ice bath was probably a good call, so we said our good-byes and parted ways.

Panic struck when we reached the car and I tried to open the door. It was locked. He'd locked the keys in the car! Then I realized that Dad simply hadn't unlocked it yet. He pressed the button on the remote, and the car unlocked. But in that instant of absentminded panic it occurred to me that for the past fourteen plus hours I'd been gone from reality, I'd been totally lost in the spell of ultrarunning, I'd been so totally immersed in the experience that nothing else seemed material—not the car remote, not the bills that needed paying, not the politicians in the White House, not the emails that needed sending; all of those things had melted away and vaporized. It was a cleansing of the soul, a physical and emotional reincarnation. After fourteen hours of running I was now someone new.

As we started our drive back to town I noticed a little wooden icon on the dashboard. Dad saw me glance at it and an awkward moment of silence followed. Of course, I recognized it and knew exactly what it was.

Dad eventually spoke.

"I thought about her a lot today."

"I did, too, Dad."

"I miss her."

"I do, too, Dad."

We were talking about my kid sister, Pary, his daughter. While we hugged a lot these days, we didn't talk much about Pary. She had been tragically killed on her eighteenth birthday, and although that was more than thirty years ago, it was not forgotten. It will never be forgotten. While I was out there running for fourteen hours, Dad was left alone on backcountry roads and in remote wilderness outposts. I would appear for a moment, and then disappear into the wild. And between those periods he would have hours to think. Hours alone with his thoughts, about life, about eighty-two years of living and the joy and pain experienced along the way.

After her death I had hours to think myself, and I turned to a bottle to help numb the pain. Things didn't go so well initially, and they were getting progressively worse. But then I discovered running and was able to run from those feelings I couldn't drink away. It saved me, but nothing saved Dad. The loss of a child is the worst infliction a parent can ever suffer. I grieved for the loss of my sister, but it was nowhere close to what my parents felt; it never could be. The worst thing that could possibly happen to a parent happened. And they were left to somehow piece together an ongoing life.

As we were driving he began to weep, and seeing him weep-ing, I began to weep. It was too much: neither of us could hold back the emotions, and we both began to sob, big, watery tears streaming down our cheeks. There was no use trying to contain ourselves; the sensations were too powerful that we had no other choice than to give ourselves over to them. We must have been quite a sight, two grown men driving down the road bawling our eyes out.

Eventually I was able to choke back my sobbing. Dad still kept sniffling, his nose continuing to run. I dug around in the zip pocket of my water bottle holster. I'd stashed some toilet paper in there in case of a trailside emergency. It was now a brown, dirt-covered, soaking wet clump, something akin to a thick dollop of chocolate pudding.

"Here," I said to him, "use this."

He looked at it, and then looked over at me, and we both spontaneously busted out in laughter, our tears and unrestrained chuckling mixing into a harmony of the human condition. And in that moment, we were both reborn. Life is cruel, wonderful, unfair, miraculous, and unbearable. It is all these things at once, but what mattered most is that we were together, that we stuck together through it all. We were doing the best we could with the cards we were dealt, but always together, faithful forever. Nothing mattered more. This was a value my family handed down to me, and a value their forebears had handed down to them.

When we got back we filled the hotel tub with ice and I slowly immersed myself. The chilly cold liquid felt refreshing, though hardly shocking. My body retained so much heat from the day that the frigid water barely registered. "I think I'll stay in here a good while," I said to Dad.

"Sounds good, Ultramarathon Man," and he shut the door.

I tilted my head back onto the ledge of the bathtub and stretched out my legs. As the water gently lapped, a serene contentment washed over me. Nothing quite compares to the inner alpenglow of a strenuous day on the trails. There were 181 runners today, and 12 of us made it to the finish of the 100K. Each of us had our trials and tribulations along the way. We slew our demons and we reckoned with some higher authority to grant safe passage. Every ultramarathon becomes its own unique odyssey in this way, an outward journey of physical conquest and an inward examination of one's true self.

When I finally emerged from the cold-water soaking, Dad had gotten dinner for us.

"That's so thoughtful, but you didn't have to do that, Pops. You've been looking after me all day."

"Hey, it's my pleasure Ultramarathon Man," he said. "But wait . . . wait . . ." He reached into the bag and pulled out two cold Bud Lights.

"Whew hew," I chortled, "let's get Dionysian!" (Yes, that was facetious—Bud Light beer is all of about 4 percent alcohol.)

We drank our beer and ate our food. It'd been a long day, though quite a resplendent one. Dad tucked himself into bed and closed his eyes. He was out in an instant. I switched on his CPAP machine and it started gurgling with his breathing, and it was the most beautiful music I had ever heard.

6

THE AFTERMATH

*That which is most difficult to endure
is most satisfying to reminisce.*

The days and nights after an ultramarathon are always a bit hazy, coherent moments followed by foggy ones, and not with any particular rhythm or predictability. Sometimes you're afforded a striking clarity of thought; at other times you can barely recall the name of your dog. Emotions swing broadly. The highs post-ultra are pure bliss—like opium, I'd imagine. But the lows can be terrible, like a hangover from a sedative mixed with cheap alcohol. Some mornings in the aftermath, crawling out of bed was tougher than running any ultramarathon; the emotional dissonance was crushing.

I was particularly susceptible to postrace bouts of self-doubt and feelings of inadequacy. Nothing I did was enough. Ever. A disparaging inner voice interminably whispered in my ear that I was a pathetic failure, a fraud. My performances were lamentable.

It was the morning after returning home from the Bishop High Sierra Ultra and it had been a sleepless night of over-thinking, mostly not good thoughts. I checked my emails: 157. I checked my status on the Western States wait list. Running another ultramarathon could reexpose me to these postrace de-mons, but I was willing to accept the consequences. It wouldn't be living any other way. Besides, plenty of my peers suffered postrace depression. Perhaps that's one of the reasons I felt so intimately connected with these people, why I loved them so. I'd moved up on the Western States wait list several positions, from number eighteen to thirteen.

Woohoo! Never mind that I was lying in bed checking my iPhone in a state of postultra gloom, I still wanted in. Although a distant possibility, Western States was seeming more real, like it could actually happen. Hope keeps dreams alive—any hope, no matter how implausible.

Pulling the covers back over my head, I resumed the roller-coaster ride with my emotions. There was a growing pile of work that needed my attention, but I lay in bed mulling over the possibilities of Western States, over and over again. I may actu-ally get in. But what if I do?

Eventually I made it out of bed to the kitchen. There I hap-pened upon my son, Nicholas. At twenty years old and home from college, he cut an athletic figure, though he wasn't a runner these days. But it hadn't always been that way.

Seven years earlier, when he was thirteen, Nicholas and I met in this same kitchen, and the conversation we had was quite unexpected.

"I want to run a marathon."

"What?"

"I want to run a marathon."

"When?"

"When I turn fourteen."

"Really? That's not far off."

"Dad, I know when my birthday is."

"Sorry, Nicholas. You surprised me. Still, you don't just run a marathon. There's this thing called *training*."

"Okay."

As a parent, one thing I've learned about kids is that you can't tell them what to do. When you do, they'll likely do the exact opposite. Consequently, I've never pushed running on my two children, fearing the proverbial parental backlash. They both dabbled in the sport, and my daughter, Alexandria, joined the running group Girls on the Run and eventually completed a 10K on her tenth birthday. She started running casually after that. Nicholas had been even more inconsistent, running as a little boy during youth soccer season, but not running at all in the past four or five years.

Thus you can imagine my astonishment the next afternoon when he walked into the house red-faced and drenched in sweat.

"Are you okay?"

"I think so."

"What were you doing."

"Training."

"Training for what?"

"For the marathon, Dad. You said I needed training."

"You do, but I didn't think you were serious. You really want to run a marathon?"

"Yes, Dad. I told you so."

I'm not sure he completely comprehended the depth of his upcoming challenge, but when you don't know what's impossible, not much is. The naïveté of youth is refreshing in this way. Still, discipline is hard work. And training for a marathon requires discipline. I remained somewhat skeptical that he'd stick with it.

The next morning I made his usual: toasted bread and jam.

"No, thanks, Dad."

"What? Why not?"

"I don't eat bread anymore."

"What are you talking about? You love toast."

"I'm Paleo."

"Since when?"

"Since I started training."

I'd followed a strict Paleo diet for the past decade. Nothing processed or refined, and no grains. The kids constantly heckled me about it. No pasta? No bread? No cake? What fun is that!

I doubted Nicholas's dedication would endure beyond missing one pizza party with his mates. But the kid amazed me. He kept to it, oftentimes returning home from a two-hour run with casual nonchalance. No big deal. I remember once, when he was a youngster, dropping Nicholas off at school after a big race. I'd labored to make it up the curb to hand him off to the teacher's assistant. As I turned around, I heard the assistant ask, "Is your dad okay?"

"Oh, he's fine," Nicholas said. "He ran 100 miles yesterday."

There was silence after that. The TA was likely thinking what a vivid imagination this young lad had. Ran 100 miles yesterday, *suuure* . . . But my kids knew me as nothing other than an ultramarathoner. I'd been doing it their entire lives, so what I did was nothing novel. I imagine they thought all daddies did this sort of thing.

Nicholas continued his training and Paleo diet all the way up to the day of the race. And this was no easy race. It was a challenging and hilly marathon on the trails of Marin County, north of San Francisco. Very little of the course was flat.

We ran together for a while. At the halfway point it got exceedingly warm and humid, and I started overheating. There was little airflow in the valleys, the thick vegetation creating a stagnant and stifling atmosphere.

"This is miserable," I said to Nicholas.

"It's not so bad."

We went on, and the hills and the heat persisted. However, he never complained during the run, not once. He never expressed self-pity or remorse about the undertaking, he just kept powering ahead. It's funny how you can spend fourteen years with someone and not know them. I saw Nicholas in a completely different light. I'm sure he was learning from the experience, but so was I. Running a marathon doesn't just build character, it also reveals it. And I liked what I saw in Nicholas. He had a certain strength of composition and a resolve I'd never seen in him before, or never noticed.

He went on to finish the marathon, and in a respectable time. He won his age group, though he was the only one in his age group. The next closest competitor was in his twenties.

Walking back to the car after a medal was placed around his neck I said to him, "Nicholas, let's run the Chicago Marathon next. The course is pancake flat; you could almost run it blindfolded."

"Naw, Dad," he replied, "checked that marathoning thing off my bucket list."

That night he went out for pizza with his friends as just another fourteen-year-old kid. He'd "checked that marathoning thing off his bucket list." One and done.

And he'd stayed true to his word, not running a day since.

Now at twenty, it was almost like having a stranger in the house again, a summertime exchange student who occasionally surfaced for brief interludes between work and social activities.

When he saw me coming into the kitchen, he asked, "How'd the race go?"

"It was good; everything went pretty well. Your grandfather says hi."

"Cool. Okay, I'm gonna go grab some chow with friends."

And off he went. That was about the extent of our dialogue these days. He didn't seem particularly interested in my running or racing and I wasn't about to bore him with some lengthy dissertation on how it went for me. Nicholas was his own man now. I knew him briefly when he was fourteen, but he was a different person now, someone less familiar to me, someone a bit reserved and closed off to dear ol' Dad.

Nicholas used to love crewing for me when he was a young boy. Many of my ultrarunning escapades were family affairs. We'd be gone for days, and he and Alexandria would stay with my mom and dad—their grandparents—in their RV, which was affectionately coined the "Mother Ship." I remembered how Nicholas used to look forward to these adventures with wild anticipation. But that was then; this was college and independence, we'd entered a different era in our stage of life together.

There's a practice I keep of always running the day after a long race. Not far, not fast, but I run. Okay, shuffle. I find that even the pedestrian movement of slowly putting one foot in front of the other for a couple of short miles serves to accelerate recovery. And frankly, the thought of running—of movement—was the main reason I'd gotten out of bed today. Work would have to wait. First I needed to unsully my mind.

The postultra comedown could be horrid. In a profound way,

there seemed little left living for. The deed was done. What more was there? Decades in, I'd dealt with these feelings of emptiness before. Sometimes crawling out of the pit came down to sheer willpower, quite literally forcing yourself to take curative measures. Ironically, for me that meant more running.

And thus after Nicholas left the house I laced up my shoes and did the same. It was a fine late spring day, not too hot and not too cold, perfect for a leisurely cruise around the neighborhood. During these slow, plodding recovery runs my mind tends to reflect on my life as a runner. I'd raced and competed on all seven continents of earth, twice over, in some of the most extreme and remote places on the planet. I'd run a marathon to the South Pole, run an ultramarathon across the Atacama Desert, run 100 miles around São Paulo with a bunch of crazy Brazilians, run all over Europe, Canada, Asia, and Latin America. I'd run for twenty-four hours on a treadmill, suspended on a two-story platform, hoisted above Times Square, on summer solstice—the longest day of the year—with the colossus screens broadcasting my every movement. A couple of years ago I'd even served as a US athlete ambassador on a sports diplomacy deployment to central Asia, running across Uzbekistan, Kyrgyzstan, and Kazakhstan along the ancient Silk Road which connects the three nations.

During my slow recovery jog, I began to reminisce about this wholly unique and unexpected envoy in central Asia running along the Silk Road. The endeavor captures perfectly the wildness and weirdness of my nomadic meanderings. Sauntering through familiar hometown turf, my mind drifted back to the Silk Road experience . . .

7

THE SILK ROAD ULTRA

Get lost sometimes, on purpose.

I talk to a fair number of people when running marathons. When you've done a couple hundred of them, conversations are bound to arise. Most are fairly inert, topical pleasantries and casual ruminations, though this isn't always the case. Sometimes the dialog deepens, veering into unexpected territory. This can get awkward, especially if I'm uncertain where the conversation is leading. Such was the case when a guy pulled up next to me during the 2015 San Francisco Marathon. Now, permit me a brief digression, if you will. The San Francisco Marathon offers an "Ultra" division, which is essentially a double marathon. The Ultra division runners start at midnight and run the marathon route in reverse—finish to start—and then run their second marathon with the traditional marathoners. I always enter the Ultra division (hey, I'm the Ultramarathon Man, right?). Thus, when a gentleman approached me at mile 6 of the traditional marathon I was at mile 32 on my second loop and had been running all night.

He introduced himself and explained that he had, quite by chance, come upon a copy of my first book, *Ultramarathon Man*, in a Bangkok bookstore. It led to an ongoing love of running. Cool. I thanked him for letting me know and picked up the pace. We were getting ready for a steep climb to cross the Golden Gate Bridge.

But he stuck by my side. He explained that he was now working with the US Department of State at an embassy in Bishkek. Come again? You know, Bishkek. The capital of Kyrgyzstan. No, actually, I didn't know that. I thought you said Bangkok.

I picked up the pace further. But he stayed with me. As the cultural affairs officer he'd been doing some thinking, you see, and 2016 marked twenty-five years of US diplomatic relations with the former Soviet republics of Kyrgyzstan, Uzbekistan, and Kazakhstan. A 525-kilometer route along the ancient Silk Road connected these three capital cities, and he was thinking that perhaps next summer I could run this route to celebrate the occasion, a sort of unconventional diplomatic envoy program, if you will. Is he making all this up as we go? How did he have any idea he'd be seeing me today with thousands of other runners? Why can't I get away from this guy?!

While I didn't know all that much about the "stans," I knew Afghanistan was somewhere in the vicinity. Not exactly a top running destination for an American. And I knew it wasn't particularly temperate during midsummer in central Asia. Asphalt-melting hot is more like it. But this guy was doing all he could. Relations between the United States and former Soviet republics of central Asia have hit a rough patch, he informed me. We needed Dean Karnazes to show that, despite our differences, our feet still pointed us in the same direction. They may be pointing us in the same direction, but mine can't point me away from this guy fast enough! I'm just a runner; getting involved in diplomatic relations between the United States and former Soviet republics was not territory I was looking to tread.

"Hey," I said, "why don't you send me an email and we can talk then." It was your standard blow-off. But he persisted. "You would do a lot to bolster US relations in the former Soviet Union."

"Yeah, send me that email. It's on my website."

"Running is the only truly international sport."

"I'll wait for that message."

"Running unites people."

Off I vanished, like fog enveloping the Golden Gate Bridge.

Flash forward to the following summer. The flight to Tashkent, the capital of Uzbekistan, was a long one, with a disorienting layover at Inchon International Airport in South Korea. Will Romine, the gentleman I'd met along the San Francisco marathon, did manage to track me down. And he proved to be quite charming (and one helluva salesman). And now I was embarking upon the Silk Road Ultra as a US athlete ambassador on a sports diplomacy delegation. All very fancy wording for a guy planning to run 525 kilometers through the capitals of three former Soviet republic nations in central Asia.

I felt like an astronaut when the plane finally touched down in Uzbekistan. Thirty-two hours in a capsule staring down at earth can do that to a guy. I'd crossed fourteen time zones.

Once we pulled into the gate and the airplane doors finally opened, the heat and humidity inside the cabin immediately became oppressive. *Here we go,* I thought.

A large black Suburban was waiting for me at the airport. This *was* the State Department, after all. The doors of the vehicle were thick and sturdy, and they made a deep, permanent sound when shut closed, like a bank vault being locked. Bulletproof, I would later learn.

The State Department's staff inside welcomed me, and we exchanged pleasantries. Briefly. Then it was straight to business. I'd received some diplomatic training Stateside, but this was more of a tactical agenda for my next few days leading up to the start of the Silk Road Ultra. It took the entire duration of the drive to the hotel to cover off on the schedule. It was going to be a busy few days.

"Oh, one last thing: your room's bugged."

That was bummer news, but I knew bedbugs were a problem in many parts of the world.

"I'm not talking about bedbugs," one of the staffers explained.

"Really? What kind of bugs are you talkin' about?"

"We've asked them repeatedly not to spy on our diplomats, but they haven't complied."

"You're kidding, right?"

"Take a look at all the smoke detectors on the ceiling of your hotel room. Half of them don't detect smoke."

"*Seriously?*"

"And when you're taking a shower, look at the areas on the mirror that don't fog over. There's a camera behind the glass."

They dropped me off. "See you tomorrow."

Walking into my hotel room was like being on a first date. It was impossible not to keep glancing up at the ceiling. *Don't stare, don't stare*, I kept telling myself. I attempted to go about unpacking as usual, trying to retain my modesty and not appear unnatural, but it was trying. There were uncomfortable moments the State Department training never prepared me for, such as I don't wear pajamas. To keep concealed, I wrapped myself in a towel and only unpeeled once under the covers. I blew a kiss toward the ceiling: *night, night my little Uzbekie comrades.*

The next day was consumed with meetings, school visits, interviews, talks with officials, presentations, ceremonies, more meetings, and more interviews. When they finally dropped me off back at my hotel that night my head was randomly bobbing up and down, like a youngster that's been kept up past his bedtime.

Once inside my hotel room I was so ready to strip down and hop in the shower. *Oh, shit*, I thought, *how could I?* This was only our second date. I was a little less guarded walking around the room now, though still a bit self-conscious. The ceiling had eyes. I wanted to shower, but the thought of getting naked in front of people I barely knew was too discomforting. Instead, I just washed my face over the sink. I had the nerve to take off my shirt, but even that felt a bit too revealing.

It was more of the same the next day: meeting after meeting, interview after interview, talk after talk, sunup till sundown, followed by a multicourse dinner that carried on late into the night. Coming back to my hotel, well after midnight, I was officially Uzbeked out. Absolutely exhausted. I just wanted to let my hair down and relax. *Ah, crap*, I remembered, opening the door to my room, *here we go again*. But this time I couldn't handle it. I'd lost all diplomatic composure. I tore off my clothes right there in front of one of those "smoke detectors" and started gyrating my hips round and round, buck naked. Whirlybird, I think they call it, right there in front of the cameras. "Ride 'em, cowboy," I hooted. "Get a load of this American!" I taunted, "He's armed, and he is dangerous!"

I could just imagine some ex-KGB guys observing me on a

grainy black-and-white screen in the basement of a nondescript government building wondering what threat level to assign my unusual behavior. Armed and dangerous: that's some weapon I'm wielding, all right!

The run started the next morning, thank goodness. I could no longer look at the hotel staff with a straight face. The US ambassador to Uzbekistan was present for the run's starting ceremony. A marathoner herself, she was very supportive of the endeavor and gave a touching speech to the crowd, wishing me luck and safe passage. Her speech concluded with a celebratory group countdown and I was sent on my way.

As it was still early in the day, temperatures were manageable. But it was a cloudless sky and later on was sure to be different—it was going to get toasty. I followed a motorcade of State Department vehicles that led the way. They would be guiding me out of Tashkent en route to the border of Kazakhstan. After about an hour of running I decided to ask the ambassador what she thought of things so far. I ran up alongside her black Lincoln limousine and tapped on the back window, in front of where she was seated. The window didn't go down. Maybe she didn't hear me. I tapped again. Still nothing. Then I noticed the front door opening slightly.

"Can I help you?" a voice emitted.

"Ah, yeah, I was just going to ask the ambassador what she thought of the run."

The voice coming from behind the slightly opened door lamented, "She can't open her window."

It suddenly struck me: this was a bulletproof vehicle, she was riding in an armored car. Either the windows wouldn't go down or they weren't taking any chances by lowering them. Inside was a US diplomat, after all.

Wait just a minute here: *I'm* a US diplomat, too, right? She's riding in a bulletproof vehicle, and I'm out here running half naked with little more than a handheld water bottle for protection. If someone comes after me what am I supposed to do, squirt him in the eye with Gatorade? Perhaps the procession of black Suburbans and Lincolns was intimidating enough that no one would dare fuck with us. Then the lead vehicle slowed. The driver glanced out the window at me, gave a salute, and abruptly peeled off to the right. The rest of the vehicles followed and within seconds the entire motorcade disappeared into a cloud of dust. I was left alone, running.

Which wasn't so bad, actually. It was the first time since arriving that I was on my own. It was a pity the trip had been so rushed, for what I'd seen of Tashkent and Uzbekistan intrigued me. The architecture was an interesting hybrid of modern and Soviet-era design, so different from anywhere I'd traveled previously. Many of the streets were tree-lined, and throughout the city there were numerous fountains and lovely parks, with hedges of leafy basil that gave off an aroma of pesto in the steamy afternoon sun. Tashkent has one of the few Mediterranean climates outside of Greece (and the area where I live in California), and because of the many rivers and tributaries running through the land, vegetation of all sorts flourishes. I wanted more time to wander freely here; it was an enchanting place.

But I needed to be at the border crossing at 10:00 a.m. sharp. That point was emphasized repeatedly during the briefings. And I, in turn, had repeatedly emphasized that running in a foreign territory was hardly an exact science: many variables come into play in these situations that could potentially delay progress. Traffic signals aren't always synchronized, I could get lost, bio breaks are necessary, and that can get tricky in populated areas. Shoes need changing. Water bottles need refilling. Sunscreen

needs reapplying. Muscles need . . . But I was politely interrupted. "Just whatever you do, don't be late."

Tashkent shrank away behind me as I proceeded farther toward Kazakhstan. The road narrowed and straightened, and
any shade from trees eventually vanished. Now I was entirely
exposed to the scorching rays of the summertime sun, and there
was not so much as a wisp of a breeze. I could see nothing before
me other than more treeless roadway and ripples of heat forming
in the distance; the border was nowhere in sight. I glanced down
at my watch: it was 9:45 a.m. I looked up again at the endless
road ahead of me. I was going to be late.

THE BORDER

"Mr. Karnazes! Mr. Karnazes! Over here! Over here!"
Border crossings are always gritty, intense places. Officials
look threatening and seem suspicious of everything, while civil-
ians appear anxious and uncertain. But the voice calling out to
me was neither.

"Quickly, quickly, we must go," she said, grabbing hold of my
arm and pulling me through the dense crowd. She was pe-
tite, nattily dressed, and flawlessly made up, a far cry from the
ruffians standing all around us. It was just 10:20 a.m., a mere
twenty minutes late, which is actually early by runners' standards, but her demeanor was rushed and frenetic. Elbowing
her way forward, we pulled up to a glass window behind which
sat a very militant-looking individual in a uniform who stared
at us with an unimpressed sneer on his face. She passed a number of papers under the window and he was slow to pick them
up, almost dismissive. Eventually he started to look at them,

twirling the edges of his mustache as he did so. Then he shook his head no.

She started screaming at him in what sounded like Russian, but he just sat there shaking his head no, no, no. She pointed to a number of diplomatic stamps and signatures on the documents, insisting he let us through. He leaned back in his chair and continued twirling the edges of his mustache, as though he were finished with us.

She kept yelling and pointing, and he kept twirling his mustache. This went on for quite some time until he abruptly sat up in his chair and with an angered look on his face turned around and pointed at the clock hanging on the wall behind him.

Then she stopped screaming and instead started gesturing at me with both her hands, as if to say, "Look at this guy." I was clad in running gear and covered with road grime. It was clear that my mode of propulsion used to get to the border had been my own two feet (and maybe slower than what she would have liked).

He turned and looked back at the clock. She said something to him in a bit more tempered tone. He thought about it for a moment and then angrily stamped the papers with his authorizations and stuffed them back out the window at us.

"*Spasiba*," I offered. I'd learned a few key words, and this one roughly translated to "thank you" (though I'm sure I mutilated the pronunciation). He gave me a slightly amused look, and I think he found the entire affair mildly entertaining, a colorful episode in the theater of life. Then he dutifully waved us along with the back of his hand. Enough of this. Next!

With papers in hand, she yanked my arm and ushered me to security, where the room was filled with a lot of serious-looking people armed with a lot of serious-looking weapons. This was no place to spout off my broken Kazakh. Still, I couldn't resist. "*Privyet*," I offered to one of the soldiers. He tilted his head slightly and raised one eyebrow, like a dog trying to interpret a human. Apparently I can't even pronounce "hello" correctly.

With a final stamping of our documents we were officially allowed entrance into Kazakhstan. That's when I was reunited with Will Romine. "KARNO," he roared, "I'm glad you made it!" I wasn't sure if he was referring to me making it as a State Department envoy in general, or making it through customs just now, which made the thought of not making it through customs just now seem all the grimmer. Whichever the reference, I was suddenly glad to be standing in Kazakhstan.

"Hey, Will, it's nice to see you again."

"Did you get a chance to meet Aigerim?"

"Ahh, not formally. We were a bit rushed."

Aigerim Begaliyeva was the head of communications at the US embassy in Kazakhstan. Dressed in an elegant chiffon and lace shift, designer sunglasses, and leather Guess flats, Aigerim looked like a prep school grad. But I'd seen her assertive side, and she wasn't to be messed with.

"Hi, Aigerim, it's nice to meet you. Thanks for getting me through customs."

She was carrying a clipboard. "We're so happy you're here. We've got a busy day; let's go over things."

She went down the list of activities on the briefing sheet as we walked out of the border crossing area. Her English was perfect, though with a detectable accent.

"Okay, are you ready to go?" she finished.

Go? I'd just run 15 miles to get to the border, and all I'd consumed along the way was water. And the border crossing itself had been a bit unnerving. I was ready to chill out and have a bite to eat.

"Ready, *here we go!*" She paraded me around a corner, and there in front of us was a huge assemblage of people, hundreds if not thousands lining the streets. There was a band set up, rows of military officials, individuals of all walks of life waving flags and holding posters with my name on them welcoming me to Kazakhstan. I was told that for many of these people this was their first time seeing an actual American. And some sight I must've been, uber-fit and dressed in running shorts, though a bit unkempt and disheveled. I'm not sure I represented your typical American.

Aigerim took the microphone and began speaking in Kazakh. The crowd went silent, and all eyes focused on her. About one in twenty words I could vaguely decipher: "*. . . sports diplomacy envoy . . . celebrating twenty-five years of diplomatic relations . . . running the ancient Silk Road . . . please welcome Dean Karnazes!*"

The crowd started cheering, applauding, and waving flags. Aigerim motioned for me to step up to the mic. Pensively I shuffled forward and took my place in front of everyone, looking dazed and dumbstruck. I moved my lips to the microphone and began to speak.

"Ahhh, hi."

There was a moment of pause, and then like a reverberating megaphone the crowd echoed back with their best Yankee accent, "HI!" Everyone was waving their flags, and the kids were giggling at this strange man before them.

I continued. "Can everyone hear me?"

There was deathly silence, not the slightest response. What was wrong? I repeated, "Can everyone hear me?" Still nothing from the crowd. Then it instantly struck me: no one understood English! Sweat seeped from my pores. What to do?!

Just then the crack of another microphone shattered the silence. It was Aigerim stepping in to save my ass. She translated

my question to the crowd and a thousand heads shook in affirmation: they could hear me.

In my speech I expressed my gratitude for the warm reception
and articulated how honored I was for the opportunity to run
across Kazakhstan. Aigerim's translation appeared flawless, and
the people seemed genuinely moved by my words. I concluded,
"We may not speak the same language, but our feet still point us
in the same direction." That last part made Will real happy.

"So let's run!" I finished.

The band started playing, confetti rained from the sky, and off
I went, joined by a procession of about two hundred individuals. People were clapping and waving those flags again, children
sprinted alongside us, a few dogs, too. I waved and smiled, high-
fiving people as we ran past, laughing and hooting, drinking in
a moment that was truly one of the highlights of my running
career. Here was an American diplomat, not lecturing from behind some podium but running down their street, engaged and
available, giving himself over to the experience, entirely devoid
of pomp and pretension. They loved it, and so did I.

Perhaps nothing in sports or in life is as accessible to all as running. It didn't matter our language, creed, or skin color, running
was a commonality we all shared. Two hundred of us ran down
that highway as one. So many things in this world divide us, rip
us apart, but here was something that united us, that brought
us together. The fact that running is available to all doesn't di-
minish its significance; it amplifies it. Everyone that day felt the
power, regardless of age, gender, or ability.

Gradually the roadway became more barren and the fanfare
subsided. Runners slowly began peeling off and eventually I was
left running alone. The mood invariably shifted when I was by
myself, I became more ponderous and engrossed. The farther

I progressed from the border, the more rustic and rural the countryside became. It was a resplendently tranquil setting and I relaxed my shoulders and ran freely, my pace orderly, footfalls striking like a metronome, the landscape taking on a dreamy, fairy-tale-like aura.

The hours passed and the miles added up. I'd now covered roughly 35 of them since starting out, and my hunger mounted. I was used to running without food during escapades of this nature—this was more often the case than not—but today was exceptionally austere. The only thing to eat in the crew vehicle were stale Kazakhstani crackers. I tried chewing on some of them but couldn't garner enough moisture in my mouth to digest the doughy bolus. It was like eating raw oats. By afternoon my hunger grew so acute I could eat a horse.

At the next interlude with the crew vehicle Aigerim prepared me for the upcoming festivities. The township of Qazygurt was a few miles ahead, and that would be our final stopping point for the day, she let me know. Being of nomadic ancestry, she said to expect a ceremony upon arrival. "They'll present you with food. Try a bit of everything to show respect."

As I approached Qazygurt it appeared that every known inhabitant had come out to greet me. The roads were lined with thousands of people on either side, waving flags and holding posters. They cheered and clapped as I ran past, and many children ran alongside me. Coming into the main square of town a band was playing, the mayor and town officials were lined up, and a group of women clad in beautifully ornate traditional costumes were holding huge platters of food to present to me.

I'd now been running for nearly 40 miles, much of it in the heat, and I was exceedingly parched, madly craving something icy and hydrating. Stepping up to the welcoming party, one of the traditionally clad women strode forward holding a round cup between her hands—the national drink of Kazakhstan, I was told. It looked like the bottom half of a coconut shell, and inside I could see a pale white liquid. She outstretched her arms and presented it to me. The entire crowd grew silent. I took the cup from her and looked around. I was on fire, the heat of the day's run concentrating inside me, beads of perspiration dripping down my face. Everyone was staring at me, waiting to see my reaction. I slowly drew the cup to my cracked lips and took a gulp of the contents, expecting something cool and refreshing. Immediately I gagged. My body lurched instinctively to expel the substance, and I fought back mightily the reflex to hurl. The entire city of Qazygurt looked on in shocked disbelief; it appeared quite obvious that their traditional drink repulsed me.

In that suspended moment of panic, a thought came to mind. I quickly drew up my hand and started waving it in front of my open mouth, signifying that the contents were hot and I needed my mouth to cool down before taking another sip. The crowd let out a collective sigh of relief. I gathered my strength and took another sip. I managed to keep it down. Then I raised my hand and flashed a thumbs-up (very American). Everyone went crazy. The band started playing and kids and parents began dancing in the streets.

I'd just consumed kumis, a traditional drink of warm fermented mare's milk. It tasted like stale champagne mixed with sour cream. Not the most desirable beverage after running 40 miles across the desert. But I kept it down. Diplomatic duty fulfilled.

Then it was on to the platters of food. There looked to be enough to feed a small suburb, like the spread you'd see at a community center luncheon. And it was all for me! There was an

exceptionally large mound of roasted meat. I immediately grav-
itated toward it. Something savory with lots of protein looked
particularly good. I stuffed a piece in my mouth. The meat had
a distinctive flavor to it, like nothing I'd tasted before. It was different, but I couldn't pinpoint what that difference was. Sweet,
almost, and nutty.

After having my fill of food, there was a presentation by the
mayor and a ceremonial dance. Then the photos began. And
continued, and continued. Everyone wanted a photo, which was
fine, except there were thousands of people and just one me. In
time taking photos became more exhausting than the running
had been.

Eventually the line dwindled and we were able to make our
exit.

"How was all that food?" Will asked.

"That was incredible. And the tastes were really different."

"Have you had horse meat before?"

"WHAT?!"

"It's a staple around here."

"WHAT?!"

"You said you could eat a horse. Whelp, there ya go."

Careful what you ask for in these parts, I guess.

That night we homesteaded with a lovely local family. They
welcomed us to their house with more food, but I'd lost my
appetite.

And thus ended a memorable day of running during the Silk
Road Ultra.

Only it didn't end. The proprietor brought out a bottle of Snow
Queen, a premium organic vodka that is the pride of Kazakhstan.
Now, I'm not much of a drinker anymore, but that shot slid down
pret-tee nicely (much more so than the kumis). And thus *properly*

ended the first day of running the Silk Road Ultra. With extremities pleasantly tingling, it was the perfect closure to the day. *Dobray nochi*, comrades. G'night.

FIVE DAYS LATER, IN KYRGYZSTAN

There's a habit I keep of thumbing through the local newspaper when traveling abroad. This isn't so much an effort to keep up on current affairs as it is an occasion to enter another culture's reality. There's something mildly cathartic in learning about events that have absolutely no significance in the place you came from. For instance, who would imagine that a missing goat could cause such a stir? But the *Hürriyet Daily News* alluded to something quite scandalous in the making.

I'd heard about the incident while having coffee on the outskirts of Bishkek, the capital of Kyrgyzstan. Now it was the afternoon, and I was edging my way closer to this capital city. And while the displaced goat story was certainly amusing, something a trite more intriguing was commanding my attention. In the air was a mild, skunk-like odor, and the entire city was swathed in a misty haze. I sniffed a couple more times. It smelled like my buddy's '68 V-dub bug back in high school. And then I spotted the culprit: the distinctive green leaf—the kind you see on the jacket of Grateful Dead albums—was an unmistakable giveaway.

"Will," I yelled at him to pull the crew car next to me, "is that what I think it is?!"

"Yup."

"Just growing wild along the roadside?"

"They don't call it weed for nothin'," he answered casually.

"Guess not; it's everywhere." I still couldn't believe it.

"They round it up periodically and burn it. But it grows back."

"Wait: Did you say they round it up and *burn* it?"

"Yup."

My olfactory system was indeed correct. Advancing farther into the haze my muscles started relaxing and my worries began to disappear. Everyone in the city seemed super mellow and happy. Men tipped their hats as I ran past; children giggled. Drivers were courteous; no one honked.

"Will," I said, "doesn't this seem strange to you?"

"Nope. I've been living here for almost a year now; nothing seems strange to me anymore."

"Really?"

"Nothin'."

The day's run concluded in the downtown area next to what appeared to be a tall department store. After finishing up with a few reporters and TV crews, Will summoned, "Follow me."

I can tell you this: that department store was unlike anything we have in America. "The best stuff's on the fifth floor," he said to me as we rode the escalator up. The most accurate way to describe this place was a massive emporium where you could buy pretty much anything on earth.

"Check this out," Will said. It was a MiG-21 Soviet fighter pilot's outfit, complete with helmet and oxygen mask.

"What an amazing replica," I marveled.

"*Replica?*" Will said with a smirk. "That ain't no replica."

"Look at this," he pointed.

It was a real Soviet cosmonaut spacesuit for sale. It looked like something straight from the set of *Iron Man*.

There was every manner of weaponry and accompanying ammo. I held up a grenade. "Is this thing real?" I asked Will.

"Let me put it this way: if you pulled the pin I'd run like hell."

I set it back down (gently).

There was a life-size stuffed grizzly bear that had to be ten feet tall. The teeth were displayed in a snarl, its long, sharp claws outstretched. Who would purchase this item, and how they would possibly get it out the door, was an entirely different topic to ponder.

There were Russian necklaces and jewels, a massive display of antique watches and imperial Fabergé eggs, there were formal gowns that looked like they could have belonged to Anna Karenina and a sundry of accompanying silk fineries, exotic caviars, and vodkas, handcuffs, brass knuckles, bejeweled and monogrammed flasks of every conceivable size and design, and all manner of attaché cases and spy equipment. And I'm just getting started.

I exited with a few choice memorabilia. I mean, who doesn't need a miniaturized surveillance monocular? Those neighbors back home better keep in order.

Once outside I took another deep breath of the pungent air. The food that night tasted exceptionally flavorful, and I slept like a baby.

8

THE LONG RUN

Endurance comes from enduring.

The Silk Road flashback helped elevate my spirits and reminded me that maybe I did have a place in this sport after all. Maybe I was leaving my own unique mark. Reaching the end of my jog, I hit the stop button on my tracking device. Three miles in about 30 minutes; hardly a world record, but well worth the investment. These unrushed recovery runs after an ultra were becoming an increasingly necessary practice as the years progressed. No longer would my body bounce back quite as speedily as my younger self. Recovery used to be an afterthought; now it was mandatory protocol for remaining resilient. And I was in this sport for the long run.

Earlier in my career I hadn't always been so conscientious about my future. There were times when my approaches to running and racing were heedless. Running back-to-back ultras with no recovery or reset in between, sleeping little, and eating atrociously. My only interest was in pushing my body as far as it could go, and the means for achieving this were overlooked. Just go hard. Every day.

Despite this imprudence, no type of overuse injury had ever been sustained, even while running ridiculous mileage and racing nearly every weekend. Back then I didn't know much better, or if I did, didn't care. Set a goal. Accomplish it. Repeat. Just get it done, whatever it takes. Thoughts about the future and ways to preserve my athleticism were absent. My older body would have to deal with that.

But now things were different. I'd come of age, and the battle was how to endure and persevere, how to extend a career doing what I loved while the clock ticked against me. Winning this conquest required fresh thinking and new approaches to resilience and self-sustainability. With characteristic fervor, I set about rethinking everything, questioning long-standing beliefs and trying new tactics. Lasting was the goal, and I would optimize myself to do just that.

Getting to this place meant being my best possible animal. What could be done to prolong myself and thrive in this new reality of becoming, dare it be said, an aging athlete. Everything in my life was reassessed and reexamined, from training to diet, to sleep and recovery, to, most importantly, interpersonal relationships. My approach was to take a comprehensive, 360-degree view of my life through the lens of being the optimal human specimen, and everything revolved around continually tweaking and improving. If my performance was off, why? What were the contributing factors that led to this, and how could they be avoided in the future?

Mastering the physical was straightforward. I had the discipline to train hard and possessed the willpower to control consumption of food and drink. And being an introvert, I had few interpersonal relationships to manage, and most of them where fairly harmonious. People that were downers were avoided, upbeat individuals courted. My personal time was fiercely guarded, no unnecessary calls or meetings scheduled, and many invitations were respectfully declined. These were tactical things, controllable.

It was the metaphysical dynamic that presented a greater test. Governing my body was the easy part; mastering my mind and spirit was anything but. Self-doubt and insecurities were frequent emotions to be reckoned with, and the turmoil and existential upheaval that these feelings could create were disconcerting. How to remain relevant in the face of Father Time? And from a practical standpoint, how to continue paying the bills and making a living from running? I'd given myself up to this sport, gone all in. That was my nature. There was no plan B, no fallback strategy that provided a soft landing. It was all or nothing. It's been said that rust never sleeps. Well then, my endurance mustn't, either. My back was against the wall, constantly. If I failed at making a living through running, there was only darkness and destitution. Perhaps this was an overdramatic assessment, but that is just how I saw things. The simple act of running is what I loved most, and I was wholly committed to remaining faithful to my true self. Anything else would be a compromise, a sin. In my soul, I was a runner, and if I were anything else my life would never be fully lived.

Working in my favor was that I loved what I did for a living and had infinite amounts of energy for pursuing my vocation. I certainly needed such stamina. There was no blueprint for making a living as a lifelong runner, no road map or career path to follow. It was largely unchartered territory, which was both nerve-racking and exciting. I got to forge my own path, to develop my own best practices and strategies for succeeding. One thing that became abundantly clear over the years is that if it were to be, it was because of me.

This was particularly true when it came to chasing sponsorship deals, which were major sources of income. Brands can be finicky creatures, and corporate objectives shift and convolute all the time. It was exhausting work securing a new sponsor, and each contract was slightly different, requiring new learnings and adaptations to new people and nuanced corporate cultures. You wanted these relationships to last, but you were at the mercy of ever-changing agendas and objectives. All the while you needed to be prospecting for the next opportunity and not allow the pipeline to dry. Landing a new sponsor was every bit as arduous as running an ultramarathon: it came down to a great measure of perspiration mixed with a dash of luck.

The partial list of sponsors I've worked with over the years would include:

Balance Bar
Compeed
Dove Men's Care
ElliptiGO
Fitbit
FloWater
Greek Gods Yogurt
Hammer Nutrition
Health Warrior
Hi-Tec
Kiehl's
MapMyRun

Marriott
McRoskey Mattress Company
Michelob Ultra
Motorola
Muscle Milk
Muse

And that's the first half of the alphabet. Then there was:

Nathan Sports
Nature's Path
Nike
Oakley
Powerade
PowerBar
Recoup Fitness
Road ID
Rodale Books
SOLE Custom Footbeds
Sprint
The North Face
Timex
Toyota
Ultima Replenisher
Volkswagen
Wild Planet Foods
Zensah
Zone Nutrition

I could go on . . .

Don't get me wrong, I am massively fortunate to be doing what I'm doing and have absolutely nothing to complain about.

My point is simply that it takes hustle and drive to forge these relationships, unremitting self-motivation. This was not glamorous work, and the consequences were severe. Have a bad year and your house gets repossessed. Worse, suffer an injury and everything disintegrates. Those are the very real consequences of the life I was playing. You had to believe in yourself, in your resourcefulness and creativity to continue making a go of it. You had to believe when others sometimes didn't.

And others weren't always kind or gentle in their actions, either. Illustrating this point was a year-end breakfast meeting I had with the new VP of marketing of a company I'd been working with for years. My contract was up for renewal, and by all measures we'd wildly exceeded our goals each and every quarter. He didn't appear to be a runner, but based on our track record of success the future together seemed bright.

Apparently his sentiments were otherwise. Not only did he inform me that my contract wasn't being renewed, he admonished me on top of it. "You need to grow up and get a job," he scoffed.

"I do?"

"Yes, you do. You waste a lot of time running. I mean, how long does it take the average person to run a marathon?"

I looked at him. "Average people don't run marathons."

He was getting ready to respond, but retracted when he thought about my answer. "Cute," he said instead.

"Just sayin'." I shrugged.

"C'mon now, let's be honest," he continued. "What do you have to show for all this running?"

"Show? To who?"

I didn't much like his question, and he could tell. "Waiter!" he abruptly harkened. He paid the bill and slithered out.

"I'll get the tip, bro," I called after him as he left. I put some money on the table and thanked the waitperson, then went for a run. I needed it. My confidence had been dealt a swift kick in the belly.

Running is a form of escapism; few runners would deny that. The metaphor of running away from one's problems is hardly allegory, and it was certainly the case for me. Though why is that such a bad thing? Having a release valve allows the buildup of toxic fumes to be vented periodically. On untold occasions I ran out the door with the weight of the world on my shoulders and in the course of 5 or 6 strenuous miles these problems somehow dissipated into the ether. Sometimes I just wanted to keep going, to leave the world behind and just run. But that would be irresponsible. Yeah, it would, which made the idea all the more appealing. Odysseus ventured to faraway lands, yet returned home to his responsibilities and familial duties in due course a renewed man. Running could be at once irresponsible and responsible in this regard, a way to escape the madness of modernity and reemerge refreshed and washed clean.

Honestly, I couldn't handle typical day-to-day living, didn't have the makeup for it. Some people lived by routine, but for me doing something predictable every day was like a slowly wilting flower, death coming in imperceptible degrees, a petal gradually falling off, then another, and another, until eventually all that remained was a shriveled and frail stem of a man.

Routine was death of the worst kind, a slow, insidious stripping of soul. Rarely could I even bring myself to run the same route on subsequent days; more rarely did I run at the same time every day. Sometimes I'd venture out first thing in the morning, other times during midday, still others in the evening or at night. I wasn't made to fit the modern industrialized world; my natural rhythms ran contrary to the nine-to-five business cycle. And I didn't always find people the preferred company. Not that I was antisocial, but being by myself wasn't unpleasant. Running alone was something I relished most my life, even more so as I'd become older. Most runners prefer to run alone, so these habits are not entirely aberrant. The world and its institutions engulf and suffocate us. We runners find our sanctuary in retreating to the roadways and trails, our sacred reprieve. The wonder isn't that we go; it's that we come back.

Our daily outings become purgings and resurrections. We move through this world as spirits, the air and the ground and the sky above absorbing us into something grander, and we disappear from the unbearable heaviness of being. These moments of transcendence cleanse our soul and liberate us from the manufactured and superficial. For a brief, beautiful instant we are as a human is meant to be, free and unencumbered, and this restores us and makes us fresh once more.

And then it's on to the follies of being a citizen, of being a useful and contributing member of society. Back to the fickleness and irrationality of human nature and the roller coaster of modern living, with its spirals and twists, letdowns and disappointments. As soon as there are people involved, things get complicated, and rarely do they go the way you want them to. Over a lifetime, nos greatly outnumber the yeses.

But the strong endure. The lessons you learn from running translate to life. The runner has a strong body and a strong heart. You get knocked down, you pick yourself back up, dust off, and keep going, only to get knocked down again, only to pick yourself back up once more and continue on, arising one time greater than toppling. And in this persistent enduring you acquire endurance. Your permanence is established in this way because you do not unseat easily, you have what it takes to withstand setbacks. You may waver and misstep, but you never give up. No matter how daunting the obstacle, you forge onward and keep chipping away until that barrier is eventually obliterated and overcome.

Nearly two and a half decades had transpired since I first ran the Western States 100-Mile Endurance Run. From that day forward my life had never been the same; something was inexorably changed deep within me. Something wonderful. My hunger to run this race now, after so many years, was just as strong as it had been the first time. I felt like an aging fighter dying to step back in the ring, longing to prove that he still had it, even after all this time. Twenty-five years later and the fire in the belly still burned white-hot. *This* is endurance.

9

CHASING WINDMILLS

Running is worthwhile in itself.

Several days passed and my status on the Western States wait list hadn't changed. Still no upward movement. I hit the refresh button three times just to be sure. Entry denied. Scarcely three weeks remained until race day, and the odds were looking more remote with each passing moment.

Why running this race again was so important to me was curious. I didn't have anything to prove, enough already. But it *did* matter to me; it mattered as much as life itself. I know, I know, that's absurd. Though I'd met CEOs who were prouder of their Boston Marathon medal than their status as chief of the company. I'd seen people's living rooms transformed into glorified trophy cases. I'd met scores of individuals with tattoos of their favorite running quotes. Few things mattered in our lives so much.

Perhaps we were all just chasing windmills. Silly games, these things. Though tell an ironman that and watch how they react. Running had taught me that the pursuit of a passion mattered more than the passion itself. If you loved basket weaving, be the best darn basket weaver you could. Pour your heart and soul into your craft and it will bring you immense fulfillment. Running was my thing. Deride me if you will, but it was my gig, and I was proud of it, blackened toenails and all. To me, running is worthwhile in itself.

Just then Julie came in. My wife of three decades, we had a certain familiarity.

"Hey, luv," she said, "watcha doing?"

"Just checking my status on the Western States site."

"Anything?"

"Nope. Still no luck."

"Okay, I'm gonna make a salad."

Honestly, her indifference was biting. We'd known each other since high school; she was my first love. Our relationship was close, perhaps too close. Familiarity breeds content, or so the saying goes. Sure, we'd changed plenty over the course of three decades, as people do, though my permutations had certainly been more sorted. I'd been a competitive surfer, had ambitions of being a professional volleyball player (at five-nine, good luck with that), paid my college tuition as a professional windsurfer, and now was a runner. I'd held jobs in health care, consumer products, high tech, organic snack foods, and bottled fruit smoothies. At six years old Julie knew she wanted to be a dentist. She'd become a dentist. And she was happy being a dentist. She'd built one of the busiest dental practices in San Francisco and thought nothing of it. All in a day's work, no biggie.

Julie had watched my journey into ultramarathoning. The

sport was nothing of interest to her, though she'd always been a willing, if not occasionally enthusiastic, supporter. But lately her attitude had chilled. Enough chasing some husband around who's gallivanting off in the wilderness at all hours. No, she'd suffered plenty a sleepless night schlepping food out to some desolate outpost for me.

That I could handle. The crew thing gets tiring after all. Julie never complained, never failed to step up when called upon, but lately I'd noticed her enthusiasm ebbing. Fair enough. Crewing is exhausting. Though it wasn't her lack of physical support that ate into me, it was her nonchalance. Like what I did was just what I did. Woo fucking hoo.

I followed her to the kitchen. "I think you're starting to take what I do for granted."

"Oh, here we go again." She rolled her eyes.

"What's that supposed to mean?"

"You know exactly what that's supposed to mean. You've been saying it to me a lot lately."

"What, that I feel underappreciated?"

"Yes, that I don't appreciate you, I take you for granted, I'm not grateful, blah, blah, blah . . ."

"Hey, is there anything wrong with a partner expressing his feelings? Can't I confide in my wife?"

"Oh, c'mon. That's unfair. You know I appreciate you."

"How do I know that?"

Just then Nicholas walked up the driveway. "Shhh," she said, scowling sternly, "I don't want to talk about this now."

"When *do* you want to talk about it?"

"Shhh!"

Nicholas walked in, and her composure changed dramatically. "Hi sweetie," she said. "How was lunch?"

"Hi Mom, hi Dad," Nick responded. "Lunch was nice, thank you. Beckett says hello."

Julie babied Nicholas, and it drove me nuts. Part of the downside of having a thriving dental practice was not spending all the time you wanted with your kids. To compensate, she smothered Nicholas (at least I thought so). The kid was in college and she still did his laundry and cleaned his room. I worried if he'd ever be able to take care of himself the way she pampered him. I reminded her of this periodically, and that made her blood boil. A mother's relationship with her son is important, I get that. But did she ever want him to be an independent, enterprising young lad? Her overnurturing was having the opposite effect.

"I'm going to work," I said, excusing myself from the kitchen.

Of course, work was right down the hallway, which was a luxury I would never take for granted. I only wish it had been this way earlier, back when the kids were younger, when I was traveling constantly and almost never home. Then perhaps I would know my son better. It's been said that in school you get the lesson and then take the test; in parenting you take the test and then get the lesson. Maybe I'd failed the test.

That night I apologized to Julie. In thirty years of marriage there are bound to be some rough patches, and we'd had our share, though we made it a point of never letting animosities steep overnight. Even the Trojans and the Greeks maintained their decorum.

The next day I checked the Western States website. And again the day after that. Still no movement. With my hope fading, so, too, I feared, was my relevance. The sport I loved was passing me by and perhaps I was trying to cling on for too long, like some washed-up jock loitering around till closing time night after night, an endless procession of last calls, of final rounds reminiscing about the good ol' days. Someone, please, put me out to pasture.

And then it happened. My status miraculously changed. I got in! Just like that, the website announced my number had come up, my entry confirmed. There were scarcely three weeks left until race day, and a lot needed to happen in that period. The countdown had begun. I couldn't be happier. The pursuit was on.

10

FRIENDSHIP AND FATHERHOOD

Putting your trust in another is
the ultimate commitment.

The next week I hooked up with a mate of mine, Karl Hoagland. An interesting study, Karl was a successful hotelier and businessman who'd discovered ultrarunning. He quit the hospitality business and bought *UltraRunning Magazine*. With no prior publishing experience, he transformed the publication into a consumer-friendly, colorful, and respected voice of the sport. He acquired the business from the godfather of ultrarunning, John Medinger (aka Trop John), who was an influential character in both our lives. I was a columnist and frequent contributor to the magazine.

We had something else in common. Karl was also running the upcoming Western States 100. This would be his tenth attempt, and he had completed all of the previous races within twenty-four hours, earning nine coveted Western States silver buckles. If he could duplicate his previous successes he would be awarded a treasured 1,000 Miles, Ten Days buckle. Like me, Karl had also crested the midcentury mark and would be fifty-two on race day. We didn't run together often, though when we did I thoroughly enjoyed our outings. He was incredibly humble, not like some successful guys that put on a humble facade, but genuinely humble, and I always learned something from him during our time together, be it in business or in life. Like me, Karl was also a father.

While other men met at restaurants or business clubs, we convened on the run. The route we chose this day was one of my favorites. While it wasn't particularly far, about 8 miles, the trail climbed nearly seventeen hundred vertical feet in that duration, not steep enough that you couldn't run the entire way, though close, especially while conversing.

Karl was sturdily built. Although a compact and solid figure, he had a boyish charm about him. Though looks can be deceiving, Karl was tough as an ox. His grittiness had gotten him through many Western States that would have halted others. During his inaugural 2005 race, he was in 363rd place at the 30-mile mark, nearly last. But somehow over the course of the next 70 miles he managed to pass 300 runners, finishing 62nd overall. Since then he'd overcome twisted ankles, massive cramps, near blackouts, and complete exhaustion to reach the finish line within twenty-four hours every time. Yes, tough as an ox he was. This year was to be his last race, he proclaimed, his crowning jewel.

However, he didn't quite seem his spry self during our run today. "I can't stop drinking beer," he professed. "Look at this belly."

"You're kidding, right?"

He grabbed his stomach. "I've gotta get rid of this."

"Mate, by US standards you're manorexic. That's hardly a beer gut."

Perhaps most interesting about his comment was not so much his desire to shed a few pounds, but his willingness to admit to a weakness. Men aren't supposed to expose vulnerability—that doesn't convey strength and invincibility. But make no mistake, Karl was a badass. Anyone who's run 100 miles in less than a day, let alone on nine separate occasions, is resilient as they come. Though running great distances had a peculiar way of whittling down the ego, of imparting a great measure of humility upon its disciples. He didn't fear owning up to a weakness, and in so doing its grip was somehow loosened.

Karl had also recently finished a tough 50K race, which likely had as much to do with his less aggressive than normal pace as any paunch. We continued motoring up the trail, volleying conversation back and forth depending on one's state of breathlessness. This is how dialogue worked during such runs: one party would express a thought and elaborate until all wind in the lungs was expired, whereupon the other party would *run* with it from there. The banter wasn't continuous and there were moments of silence where the only audible resonance was that of footsteps and deep inhalations. Sometimes it was impossible to talk, and it was just two men with their heads down working as hard as they could. Those moments were golden.

Karl seemed more pragmatic about his running than I was. He had a goal to achieve—finishing Western States for the tenth time—and training was part of the equation for accomplishing that task. At times I held a similar mind-set, though mostly I ran for something less utilitarian. Running was uplifting to me, a natural mood enhancer and innately vitalizing. Something inexplicable clicked within. Like a good song, the words didn't always make sense, but you liked the way it made you feel. Running gave me a warm internal sensation. Racing was more of a concert experience, rocking with the masses, still enjoyable, just a bit less soul stirring than running through nature with no particular purpose other than running through nature.

On the downhill segment of our jaunt I asked Karl about the possibility of doing a long training run over the upcoming weekend.

"Sorry," he said, "can't. Kid duty. Hopefully the money's in the bank."

That last comment was colloquial runner-speak—training put money in the bank, and racing drew down those deposits. With just two weeks to go before Western States, my bank account was looking anemic.

We concluded our run with a fist bump, and then Karl peeled off down one split in the roadway and me down the other. "See you in Squaw," I offered in farewell. It was more colloquial, this time of the ultrarunner genre. The start of the Western States 100-Mile Endurance Run was at the base of the Squaw Valley ski resort.* #SeeYouInSquaw was a hashtag and an inside refrain among ultramarathoners. "See you in Squaw," he said in parting.

Our home was on a hill, which meant that every run from the house concluded with an arduous climb back to it. And my

---

* The name Squaw Valley will be changed in 2021. The new name will not be announced until after the publication of this book, but I thought it important to recognize this name change.

personal policy was never to drive to the start of a run, which essentially meant that absent of any unforeseen circumstances every run started out my front door (besides, I don't own a car, which is a natural policy enforcer).

And that, too, the kids thought strange. What father didn't own a car? If I wanted to get anywhere I relied on self-propulsion, and I encouraged them to do the same (okay, fair enough, as young kids they didn't have much choice in the matter). Nicholas got his first bike at age nine, and we used to commute the mile and a half to his school, me running, he riding. An afterschool game we'd occasionally play was to race home. The approach leading up to the final hill our house was perched on was relatively mild, and Nicholas would rocket ahead of me on his bike. But as the severity of the climb steepened he was forced to slow down, where I could erode his lead and catch him. He tried diligently to stay in front of me, but I was older, bigger, and stronger, and inevitably I would pass him in the last little bit and beat him to the house. Turning to him during one such contest, I looked over my shoulder as I passed and quipped, "You may beat me one day, but this is not that day." He was red-faced, huffing and puffing, and trying assiduously to stay ahead of me. But it was no use.

When we got home Nicholas didn't say a word, he just looked at me. He had these naturally sad eyes and I couldn't tell whether his eyes were sadder than normal. He just looked at me, didn't say a word. It sent shivers down my spine.

Even today, many years later, I still can't get those haunting eyes out of my head. Did I screw up? The jest about beating him was meant in fun, but did Nicholas interpret it as such? I don't know how it made him feel, no words were exchanged, he just stared at me with those big, penetrating brown eyes of his. Had my words incised wounds upon the very child I loved? Did my actions as a father inflict lifelong scars?

I'd never been much of a "Let's play catch, son" sort of dad. If Nicholas asked me to play catch I'd abide, but such invitations were rare. I attended all of his games when I wasn't traveling, which was about half of them, not that it seemed to matter much to him. He maintained a casual apathy on the subject, as if he really didn't care one way or the other. Though occasionally I'd catch him glancing over at the sidelines to see if Dad was watching, which made me believe that perhaps my presence mattered to him more than he let on.

Years later, in high school, Nicholas became a standout football player. I went to these games as well, which, truthfully, wasn't easy for me. What made it so hard to watch was the fact that these were big, powerful kids on the playfield, and when one charging helmet got lowered toward another I could see nothing good coming from it. The crack of helmet-on-helmet impact was a distressing bolt of concern to every parent. I had to look the other way on many instances. Still, I went and watched.

Nicholas rarely commented on my presence, or lack thereof. The family had endured a rebellious stage with Alexandria, which wasn't easy. She could unleash a torrent of venomous pronouncements that were enough to cut any parent to their knees. Angry shouts were sometimes uttered; doors slammed. Still, with Alexandria we knew what we were up against—she was vocal in expressing her angst. Nicholas kept it all inside. I often wondered if the top would blow off one day. It wasn't healthy repressing one's feelings like that, but there was no conveying this to Nicholas—he would only retreat deeper within himself. He didn't talk to anyone.

When I got back to the house from my run with Karl, Nicholas had just returned from his summer job.

"Hey, Pops, how's it going?" he greeted me.

"Just fine, Nick. How's the new job?"

"It's okay. Did you get into Western States?"

Honestly, I didn't even know he was tracking on it.

"Ironic you should ask. Yes, I just got in."

"Cool. I'll crew for you."

"Really?"

"Yeah, sure. Hey, I'm gonna meet some friends. See you later." And off he went. Just like that, he was going to crew for me.

That night I told Julie about our conversion. "How wonderful," she said.

"Are you kidding? He has no idea what he's getting himself into."

"He's crewed for you before."

"Julie, the last time he crewed for me he was eleven years old."

"He'll do fine."

I wasn't so sure. Nicholas was an enigma to me. He complained that his college roommates were slobs, but every time I glanced in his bedroom it looked like a cyclone had just blown through. Julie still cleaned up after him, still made his meals. I wasn't sure he was capable of taking care of himself, let alone me. And your crew is your lifeline during an ultramarathon. Having what you need when you need it is critical. Conversely, not having what you need when you need it can be disastrous. Nicholas had been a very young boy the last time he'd crewed for me at Western States, and Julie and my friends Kim and Topher Gaylord had done most of the heavy lifting. Nicholas was just along for the ride.

Could he get the job done now? Was he responsible enough,

and capable? I wasn't so sure. But I'll tell you what: I wasn't about to say anything. Now it was my turn to keep it inside. For all my many faults and transgressions as a young parent, I wasn't about to repeat those same mistakes twice. Putting your trust in another is the ultimate commitment, and I would put my faith in Nicholas. Could he pull it off? We'd have our answer in a few short weeks.

11

LOST IN A WHITE HOUSE

Your feet can take you to
some interesting places.

My wife's cure for everything is fresh air. Not feeling well? Open the windows. Don't like the news today? Let in some fresh air. Room feeling stuffy? Air the place out (and the room was *always* feeling stuffy to her). Consequentially we lived in a barn. And as a runner I have no body fat, nothing to insulate against the savage elements that are my living room. Luckily, one of my sponsors is The North Face, and a perk we athletes receive is a modest stipend to purchase clothing for our personal use. My closet is filled with big puffy jackets. I need these to survive a night at our house. Even in midsummer.

There are some battles with Julie that just aren't worth fighting. You'd suffered enough defeats in the past to know that any attempt was futile. No, save your energy for writing personalized handwritten thank-you cards to anyone that so much as gives you a fruitcake for Christmas, even if you didn't like fruitcake. *Especially* if you didn't like fruitcake. Demonstrates character.

She has her ways, but mostly she's an angel. I get all the credit for being this studly indefatigable ultramarathoner, but let's be honest, she does all the hard work. Get a sick kid to the doctor. Done. Ready a meal from scratch in fifteen minutes. Presto. Help with homework—even chemistry—no problem. Recognize every employee's birthday with a special luncheon celebration. Goes without saying. And always with a smile and a wink. I possessed half her endurance and a tenth of her charm. She did all the work, and I got all the recognition.

It would be easy for someone in her position to express jealousy and misgivings, but Julie never did, and when I told her I was going to quit my corporate job and try to make a go of it as an ultramarathoner she looked at me and said, "I wondered what took you so long."

Thus, a few years ago when I told her of my intention to spend a couple of months running across America she responded with her characteristic positivity: "How exciting." Not, what about me? Or, who's going to take care of the kids? Walk the dog? Take out the trash? Help with the laundry? Just *"How exciting."*

Of course, the thought of running 40 to 50 miles a day for seventy-five days straight wasn't everybody's idea of excitement, but it was mine, and Julie understood this. And she also understood herself. While I delighted in the idea of a coast-to-coast leg-powered migration, she was a nester. Spending seventy-five days solo in the roost wasn't entirely unpleasant.

Relationships can be complicated enterprises. Everything's just fine until another person is involved. And introducing ultrarunning into the fold is like bringing home a beehive. You either learned to live with the new threat—perhaps even enjoying the sweetness of the honey—or you got stung so severely you ran out the door screaming. I'd seen relationships grow closer when one partner begins roving excessively, and I'd seen relationships blown to smithereens. Things inevitably go one way or the other; there is no copacetic middle ground.

Our companionship had stabilized over the miles and we'd struck a harmonious accord. I'd now been gone from the house for more than two months running across America and I missed Julie and the kids immensely. But the longing seemed to strengthen my love for them, and I hoped theirs for me. The truism that distance makes the heart grow fonder seemed genuine; they occupied me in some way every day and every moment during my transcontinental foot crossing.

In crossing America, starting on the West Coast and heading east works best. Trust me, I've done it both ways. Doing it in the opposite direction means running into a nearly constant headwind. No fun. Thus I'd started in LA and was heading for NYC, west to east. I'd left California sixty-five days earlier and was following mostly backcountry roads and pastoral byways whenever possible. There are multiple paths one can follow when traversing the country, though most of them inevitably funnel you through Washington, DC, which is precisely where I found myself on this day. Coming into the capital corridor, I'd been running nearly two marathons a day without rest and was admittedly a bit gruff and unkempt. I'd let my appearance go; that simply wasn't a priority. Who would recognize me out here anyway?

During yesterday's run I'd gotten a funny message from my crew, so farfetched that I'd written it off as a joke. "Really?" I saw through the gag. "That's preposterous," and back to running

I went. They were obviously bored, making things up between handing food out the window. But they said to expect the unexpected.

Today the weather was clear, a bit on the warm side, but the air was remarkably fresh. I'd begun the trek in late February and it was now April. Spring had sprung. I, too, had sprung. Not showering for days on end can leave a man with a certain bison-like muskiness. Eventually I needed to find a hose and rinse off, this much I knew.

Patriotically, the path I was on led me straight down Pennsylvania Avenue, home, of course, to the White House. After traversing the Great Plains and running over mountains and across deserts, it was somewhat disorienting to suddenly find myself ambling down America's most renowned street. But here I was, running past that whitest of houses.

There seemed to be lots of commotion going on today, what appeared to be confetti on the ground and groups of tired, lingering revelers, as though there'd been some sort of celebration. Hmm . . . I don't recall Pennsylvania Avenue looking like this in the past, but perhaps times have changed. There had definitely been a party last night.

The austere white pillars of the White House came slowly into view as I sauntered down the pathway, perspiring a bit and feeling the grunge of two months of running mixed together with bodily moisture; a sort of kefir-like paste had permanently formed in the folds of my armpits and other sun-deprived crevices. Some of the tourists glanced guardedly, perhaps concerned for their safety. Rightly so. The grunting and unintelligible murmurs coming out of my mouth weren't bettering my appearance. It'd been a long, long traipse across the country.

Suddenly the gates of the While House began to open. The

first thing I saw was this futuristic-looking firearm, like something the Terminator would carry. Was my ex-governor visiting? (Remember, I live in California.) *Steer clear of this one*, I thought, angling toward the opposite side of the street. Most of the tourists stood transfixed.

Then the most unexpected thing happened: the man holding this weapon motioned to me. Was I under arrest? I wanted to run for it, though swift as my legs may carry me, there's no way I'm outrunning a laser-guided projectile from that bazooka he's gripping. I tried to keep to myself. But then it happened again: the man at the gate motioned to me. This time I obeyed his request and tepidly meandered toward him.

He cut a foreboding figure, muscles bulging and a jawline of geometric angles. As I approached, he took his left hand—the one not holding the weapon—and held it in the air. Was he indicating for me to halt? Is this when I get cuffed and carted away?

"KARNO," he roared, "good to see you, man! Come on in!" He nodded to proceed, and then gave me a high five with his outstretched hand as I vacantly stepped through the gateway. "Straight that way," he instructed. Nearby observers couldn't believe what they were seeing. The gate closed behind me. I was now inside the White House compound.

Once inside the compound, an individual standing at the doorway of the White House motioned to me, *C'mon, c'mon, right this way*. Pensively I jogged toward her, still in a state of suspended disbelief. Is this when I wake up in a delusional cold sweat? "Right this way," she said with a wave. "Just head down the hallway."

I followed her directive and ran into the house. As I ran through the doorway it smelled regal and stately, like an old library. There were portraits of past presidents lining the way, colorful flags, and busts of prominent men in well-lighted alcoves. There were old books piled high on sturdy desks of mahogany and pine commissioned by our nation's forefathers. I was like a young child running down his hallway, only this was the White House. Ornate chandeliers dangled overhead. I ran past more statues and significant pieces of presidential memorabilia mounted in glass cases. Staffers were cheering and clapping as I passed. I noticed them through my peripheral vision, they were in fancy suits and well groomed. I smelled expensive men's cologne. I, on the other hand, smelled like a woolly mammoth. The carpet was red, accented with gold stripes on either side. It was plush to run on, soft, undoubtedly of the finest quality. My shoes were covered in road grime and dust. I was surely leaving footprints. Would taxpayer money be used to cover repairs? I felt guilty.

Eventually I reached an important-looking individual. She was holding a clipboard and had lots of credentials and security clearance badges hanging around her neck. What happens now?

"Hang a left," she instructed. "She's waiting for you out there."

Out there just so happened to be the South Lawn of the White House, and as I exited the building I saw a figure standing in the distance. It was bright outside, and my pupils needed time to adjust. I continued running toward this indistinct figurine, my eyesight gradually coming into focus. As I drew nearer, the form crystallized.

It was Michelle Obama, the first lady.

HOLY CRAP! was my initial reaction. Then a wave of panic gripped me. Did I just mouth that? I was sure there were sophisticated facial recognition cameras trained on me, watching my every micromovement and gesticulation. It was all happening too quickly.

Though my next thought was a bit more sobering: How do you properly greet the first lady? Do you bow to her? No, you idiot, she's not the queen. Do you kiss her hand? Unlikely, she's not Mother Teresa. What's the right thing to do? My mind was racing, my feet were moving, and I was getting closer, heart pounding wildly. *Think, Karno, THINK!* Nobody trains you for shit like this.

But the first lady solved the conundrum by opening her arms! I think she wants to hug you, I try to process, I think she wants to embrace. Wait, did I put on deodorant this morning? Dude, you haven't put on deodorant in the past two months! You haven't showered, either. And can you even remember the last time you brushed your teeth? Hang on, your wife's a dentist, you brush and floss twice daily. Do I actually hug the first lady? Do I dare? *Think! . . . THINK!*

It's too late. We are embracing. I am hugging the first lady on the South Lawn of the White House. We are locked in an embrace. She is wearing a beautiful silk blouse; I am robed in crusty running gear. *This isn't happening,* I tell myself. *Stuff like this doesn't really happen. Wake up.*

My thoughts are interrupted. "Dean," she says, "it's such an honor to meet you!"

Hold it, did she really just say that?

But it gets even better. "The girls and I have been following your journey the entire way."

Girls? Like, Malia and Sasha, your daughters? The Obama family is watching me run across America on TV? And I'm not showering? Suddenly I feel very un-American. I'm going to start showering more often, I tell myself, and putting on Ralph Lauren aftershave.

A reception follows with a talk to some local schoolchildren about the importance of regular exercise and physical activity. "Be just like Dean," First Lady Obama instructs them.

Umm . . . just like Dean? You sure 'bout that? You might encourage them to take up occasional jogging, or perhaps join the school track team. But be just like Dean? That's a bit extreme, wouldn't you say? (Though I don't correct her.)

Her talk concludes and a fleet of mini presidential golf carts comes wheeling out. They are filled with healthy and wholesome Michelle Obama–approved snacks. The kids seem disinterested, though, like pizza would have been better. Instead, they start chasing each other around on the grass in giddy delight. Who could blame them? As far as schoolyard playgrounds go, it doesn't get much better. Hell, if the first lady wasn't standing right next to me I'd be running around like a child, too. But she is standing right next to me. And boy, is she tall. I'd say more than six feet. And she's wearing high heels. Having run so many miles, I'm short (and undoubtedly getting shorter). "Do you mind if I take off my shoes?" she asks.

Wrap your head around that one. The first lady is asking me for permission to take off her shoes, asking if it's cool for her to kick around barefoot on the South Lawn? I do not contemplate extensively; I bob my head in consent. We take a family photo. Michelle Obama is barefoot.

And then, as if saving the best for last, she announces, "I have a special guest I want you to meet."

Special guest? Could it be, I gasp, THE MAN?

"I'd like to introduce you to Bo, our family dog."

Bo comes bounding out, not exactly THE MAN I was hoping for. A Portuguese water dog, Bo is incredibly well behaved and impeccably groomed. I detect no fleas on this dog and he does not attempt to sniff my crotch. He smells better than I do. We run around together. He does not bite, or even nip. I am certain he has been properly educated, probably at an Ivy League doggy day care center. Runners like these kinds of dogs. Bo is much more polite than those vicious mongrels I encountered on the back roads of the Ozark Mountains. Those mangy scoundrels could use a lesson in conflict resolution from Bo. "Good dog." I can pet Bo without losing fingers. Yes, Bo could teach his unrefined brethren anger management and help them develop strategies for treating passersby on foot with greater kindness.

Bo licks me—civilly, of course. The entire scene is beautiful, like a Norman Rockwell painting. I am playing with the Obamas' dog on the South Lawn of the White House. Michelle is walking around barefoot (we're on a first-name basis now). She is munching an organic apple. The sun is shining. My family is here. It is a perfect moment, supremely flawless in every regard. And then I feel someone tugging on my coattails. "Karno, we've gotta split. You still have 25 miles left to cover."

It is one of my crew. They're informing me that I must leave. "But it is the perfect moment," I plead. They'll have none of it. Reluctantly, I bid the first lady adieu and am escorted to the exit. We depart, the gate slamming shut behind us. I prepare to start running again. But before I do, I have a question. I turn to my escort. "Tell me, what was all that commotion last night on Pennsylvania Avenue about? All that confetti? Was there a party?"

"You don't follow the news very closely, do you?"

"Ah . . . not for the past sixty-five days. I've been a bit preoccupied, the running and all. What happened?"

"They got Osama bin Laden last night. A group of SEAL Team Six finally took him out. There was a huge party in front of the White House. I think you were supposed to meet the president today, but he's been on lockdown in the Oval Office. So I guess you got the dog instead."

The dog instead. Perhaps one day I'll tell my grandkids about the time I visited the White House and shook the paw of the presidential dog. And they'll giggle, silly grandpa, and go back to their video games.

Silly grandpa, that is how I will be remembered. In the world of ultrarunning, being invited to the White House doesn't count for much, it was merely an obscure personal artifact that I clung to in response to my inability to find solid footing in a sport that was passing me by. This is what happens when you grow irrelevant, I suppose, you start reflecting backward instead of moving forward. You find yourself in the garage dusting off old newspaper clippings trying to relive those bygone days of glory.

In a couple of short weeks I would have a reckoning with a 100-mile trail that couldn't give a rat's ass whether I'd shaken hands with the president or the pope. My reputation and prior deeds meant nothing; judgment would be based solely on one thing, my ability to make it from the starting line to the finish line. And that thought terrified me.

12

JUST DID IT

When you get to where you're going, keep going.

It was doubtful that my pancreas and spleen were actually being cleansed, but the instructor proclaimed these—along with a myriad of other health benefits—would be ours if we could just continue twisting our sweat-drenched limbs into a constrictive human knot and hold it there for a tad longer. Inside the studio it was stifling, the room heated well north of a hundred degrees, the pungent aromas of perspiring humans suffusing in the dense and dormant air. Her coaching came in loquacious gushes of encouragement. "You can do this! Hold that position! Little sips of air . . . breeeathe."

Tied into a human figure eight, I thought I could be losing conciseness. But she paraded about the studio, quick to point out noncompliant yogis, her headset amplifying the admonishments. "Don't fall out of posture! Tighten the core. Use your arm strength. Breeeatheee." My extremities were turning purple but I wouldn't dare untangle myself, the communal peer pressure being far too intense, mirrors on every wall for ready scrutiny. And that's what I like about yoga: it's so nonjudgmental.

Soon we would be able to release the pose, but her countdown from ten was periodically interrupted with more garrulous declarations. *"Nine . . . eight . . . you are the light of the universe, feel the glow inside."* A silver stud pierced her tongue, and her belly was tattooed with what appeared to be a lotus flower blossoming out of an exploding cassava melon. I tried not to stare. *Find your center, Karno,* I told myself. *"Seven . . . six . . . inhale the future, exhale the past."* Her hair was orange. *"Five . . . four . . . letting go is the hardest asana, do not cling."* Beads of perspiration dripped from the tip of my nose. *"Three . . . two . . . become your inner peace . . ."* The countdown was the only time she talked slowly, ten seconds dragging into infinity, the heat amplifying and suffocating more entirely with each passing millisecond.

Finally she resumed, *"One and three quarters . . . one and a half . . . one and one quarter . . . releeeasse . . ."* I struggled to untie myself, fearing permanent neuromuscular damage.

And to think, I paid to be here.

It was a week before Western States, and hot yoga was good preparation. After ninety minutes of this ungodly contortion our session finally concluded. Some people lay idle on their mats, palms facing skyward, grateful for their success. Or dead. It was hard to tell.

Personally, I don't like hanging out in a room where a mass of humanity has just spent a good deal of time excreting every manner of bodily toxin. The instructor promised these things were being released, but better not to stick around to test the theory. Nope, I couldn't get to the exit quickly enough.

"Namaste," she said while bowing to me on my way out. "Na-maste," I said, bowing back in gratitude. Even my lymph nodes felt revived, I wanted to tell her, as she assured us they would.

"You did better today. Remember, yoga is good for running, but running is not good for yoga."

She told me this every time I took her class. Still, I didn't say anything. I really wish I could tie myself like a human shoelace, but there was a modicum of truth in her saying. Runners are notoriously unlimber, myself being no exception. Yes, I can touch the ground from a standing position, even with my palms, but scratching my forehead with my toes, forget about it.

These sessions of hot yoga were beneficial, despite running 2 miles each way to and from the studio. The yoga instructor would just as well have me stop running, but that wasn't going to happen. Instead, I did what I could to counterbalance the toll running took on the body, like taking extended sessions of hot yoga. Truthfully, I wasn't convinced running was all that bad. I didn't buy into the narrative that the human body has a finite number of footfalls and once you hit that mark the system deconstructs. I'd seen too many people in their seventies, eighties, and even nineties finishing marathons. Compared to their peers, to me these folks looked to be in terrific shape.

One of the reasons I logged so many cumulative miles was that I didn't own a car. At first the transition was challenging, mostly because I'd become hideously reliant on getting anyplace I wanted quickly. New thinking was required to slow the pace. Attending a meeting in San Francisco from our house in Marin County necessitated a couple hours of lead time. This took forethought and planning, that's all. Shopping trips to the grocery store became more frequent, given there was only so much food I could fit in my backpack. Though the upside was that the food we were eating was fresher. Eventually I appreciated life without a car and it felt quite natural. The constraints placed on me slowed life's pace, and that didn't seem like such a bad thing. The toughest element about not having a car was living in a world where everyone else does. They moved so quickly while I had downshifted. Though when I peered more closely, it sometimes appeared it was *they* that looked the worse for wear, not this slowpoke in his running shoes. I was doing just fine at this mellower pace.

Our actions in life ultimately shape who we are, though perhaps our inactions exert more of an everlasting influence. Many people are not happy with what they do, yet possibly worse, many people simply tolerate what they do and never take the initiative to do anything about it. Either out of fear, complacency, or sheer exhaustion, they go through the motions day after day of living a life that is less than what they'd hoped. Inaction becomes permanent, and suddenly it's too late. Perhaps the only tragedy approaching that of a young life cut short is a long life left unlived.

I saw this eerie contented discontentment all around me when I was younger working at a large corporation, and it scared the hell out of me. People were simply showing up to work and making it through to the end of the day, and every two weeks collecting a paycheck. I didn't know the definition of success, but this didn't seem like it to me. Even a life that was a failure seemed better than a life that was empty. Sure, to dare is terrifying, though the alternative was something worse. Dying, to me, seemed like a better alternative than not fully living.

Thus I decided to navigate my own course through life. This is what a nonconformist does, really. Those who set out on their own can't follow a path less traveled, because there is no path to follow. It's largely orienteering without a compass, sometimes finding a clearing that makes the journey more effortless, but mostly bushwhacking through the thicket. Everyone's unique calling was different; there were no preset GPS coordinates.

That's not to say there weren't people I looked to as guide-posts. Tony Hawk was one such individual. There was no script for forging a living as a professional skateboarder; Tony made things up as he went along, and in doing so he created success for himself and a blueprint for others to follow. Laird Hamilton was another. Professional surfing existed when Laird was coming of age, but that wasn't the kind of surfing he enjoyed. Many compe-titions were held in less than ideal conditions and were tailored for spectators to attend. Laird preferred chasing the biggest and most savage waves the planet could dole out in oftentimes far-off locations, and he turned big wave surfing into a lifestyle and a living. Then there was Conrad Anker, whom I met in my early years with the outdoor gear and equipment maker The North Face. "Rad," as he was known, hung around a cadre of so-called dirt bag climbers who largely lived in vans or on a buddy's couch when they weren't climbing mountains. Conrad stayed true to his passion and continued perusing the "raddest" global expe-ditions imaginable, making a name for himself, not just within the hard-core climbing community but also within the general populace. People lived vicariously through his adventures, my-self included. There seemed no challenge he wouldn't take up. His sense of adventure is illustrated when I invited him to run a 50K ultramarathon (which ended up being more like 55K). He took up the offer and with little training finished the race, and in a respectable time. But what surprised me most was that he looked like he could keep going.

If there was one observable quality all of these guys seemed to possess it was a genuine passion for their chosen pursuit. Be it skating, surfing, or climbing, love and passion guided their actions and behaviors, and the careers they built came as a by-product. They stayed true to the one thing that made them feel most alive, and all necessary details fell into place on the back end.

Of course, this is never quite as easy as it sounds. Sustaining oneself for multiple decades in a self-constructed livelihood is anything but effortless handiwork. That youthful vigor and enthusiasm that once provided the boundless energy necessary to tackle anything dwindles over time. A once reliable tailwind weakens and occasionally eddies against you. Weariness becomes an unwelcome bedfellow; the hard knocks begin taking their toll, the tough experiences, comedowns, and disappointments adding up even as the achievements and accomplishments do the same. How long can this go on? I often wondered. Sometimes I would lie in bed at night staring into the darkness, worrying about the future. My trophy room was filled with hundreds of awards and medals from some of the toughest conquests on earth. Hundreds of them, maybe thousands. *Nike* taunts us to "Just Do It." And I had. But what happens after you've just done it? They don't have a catchy slogan for what to do next.

That evening I bumped into Nicholas back at the house.

"Hey, Dad."

"Hi, Nick." He'd just gotten out of the shower and smelled of Paco Rabanne cologne. "Do you want to talk about Western States? It's next weekend."

"Mom's taking me over to Michael's."

"You can take the car if you want."

"We're gonna have some beers so I don't want to drive."

"Nicholas, you're twenty years old."

He looked at me incredulously, as though he had no idea the

point I was making. The drinking age in California is twenty-one, but that didn't seem to register with him.

"So when do you want to talk about Western States?"

"Just let me know what you want me to do."

"It's a bit more complicated than that."

"Dad, I've crewed for you before."

"Nicholas, you'd barely crested double digits the last time you crewed for me. You were a little kid."

"I remember. It will come back to me."

"Nicholas, this is important."

"I know, Dad." His phone pinged. "I gotta go."

Off he dashed, the room still smelling of his cologne.

That night I lay in bed worrying. I worried about sustaining myself as an athlete, I worried about how I was going to pay the bills, I worried about the environment and the world we were leaving our kids, but mostly I worried that I'd set my son up for failure. Nicholas hardly seemed invested in crewing for me. Perhaps I was overthinking things, but I interpreted his apparent disinterest in supporting me as a consequence of his perceived shortfalls in my efforts supporting him as a father. I was never there for him, globetrotting and aloof, off on some ultramarathon in who knows where.

The Western States 100-Mile Endurance Run was next weekend, and I felt physically prepared. But you don't run Western States with your body. Was my mind up for the challenge? I lay in bed churning that question over until sunrise. What would the future hold?

13

THE CAVS AND THE CAVS-NOTS

All racers are runners, but not
all runners are racers.

Tapering is not a skill I'm proficient in. This practice involves progressively lightening your workloads in the days leading up to an event, and two hands would be needed to count the number of coaches that have lectured me on the importance of dialing back intensity and distance before a big race. Perhaps it is my ears that need repair; I've never seemed to hear the message.

It's hard to say if my past performances would have been better had I followed the tried-and-true formula of a professional coach instead of haphazardly going about things on my own. Proper tapering was not something I took seriously, though perhaps if I had I would have fared better. Honestly, I didn't care all that much about my results, I was in it more for the experience, and to me turning Western States into a race somehow demeaned its sacredness. Back in the day I could reliably finish in the top ten and receive automatic entry into the following year's race, no lottery. Finish top ten; that's all that mattered. I didn't care if I won, I just wanted to run the race again. I loved it so.

Inside the Palestra, the storied Philadelphia gym known as the Cathedral of College Basketball, there hangs a banner:

To win the game is great. . . . To play the game is greater. . . . But to love the game is greatest of all.

Beating others was never a primary motivation. Conversely, being beaten never troubled me much, either. I ran ultras because I loved the game. Everything about it was glorious—rising before dawn, feeling the sun on your shoulders, being completely absorbed in nature all day, observing as the sun dreamily set and the moon unhurriedly emerged, the sky slowly filling with a million twinkling stars. Oh, how I loved it so.

Perhaps this is what kept me going for so many years, never fatiguing, never burning out. There weren't many that lasted for decades in this sport. Either through injury or waning passion, they faded away. The person I admired most in the sport, Ann Trason, had an illustrious career, winning nearly every race she entered, often outright. If Ann Trason was in a race, the contest was for second place. At Western States she was unstoppable, winning every time she entered, usually by hours.

I didn't know Ann, but I held her on a lofty pedestal. During the 2003 Western States I happened to come upon her with about 5 miles left in the race. It was after midnight and both of us were running with headlamps. When she heard footsteps behind her and saw the light of my headlamp, she yelled, "WHO IS THAT?!"

Taken aback, I responded, "Ah, it's me. My name is Dean."

Ann continued, "WHERE IS SHE?!"

Her screaming alarmed me; I thought something might be wrong. And I didn't understand her question. "Ah, I'm not sure what you're asking."

"EMMA! HOW FAR BACK IS EMMA?!"

As I scooted around Ann on that narrow path through the mountains it occurred to me that all she cared about was winning the race. She didn't see the magnificent vistas along the course, or any towering pine trees, or never mind the roaring American River, or the brilliantly vivid sunset; nor was she particularly interested in me. All Ann wanted to know was how far Emma Davies was behind her.

It was a dog-eat-dog world out there. Ruthless and cold-blooded competition could bring out the worst in people. Ann beat Emma that year, though this time it was close, the win narrow. Ann Trason never ran Western States again. One of the greatest competitors the sport had ever known simply vanished.

Ultrarunning represented to me something that had the power to rise above this insanity. An ultramarathon was a way of transcending the heaviness of everyday living. Turning a 100-mile journey through the mountains into a competition against another human seemed to defeat the very reason for doing it. We run through the wilderness because it changes us; to be fraught about potentially losing to another competitor somewhat spoils that wonder. Perhaps these were just my own romantic inclinations, but turning an ultramarathon solely into a competition made it seem more ordinary. And I was drawn to this sport because it was anything but.

If you think about it, there's a distinction between running and racing. Racing involves a gun going off, a predetermined course to follow, and a dash to the finish line as quickly as possible. The racing mentality is about beating the competition, about being better than the rest and standing highest on the podium. There's nothing wrong with this; it just wasn't my primary reason for running.

And apparently I'm not alone. While something like nine million individuals finished a footrace in North America in 2019, five times that many people in North America identify themselves as runners. They run for the experience, not the medal. All racers are runners, but not all runners are racers.

Those in the competitive world of running (i.e., the *serious* runners) tend to fixate on racing. Sometimes we forget that running can be something entirely unrelated to time and speed. I remember once telling a taxi driver that Eliud Kipchoge had just broken the two-hour marathon barrier. "That's amazing!" he replied. "How far is a marathon?"

To me, running was about more than speed and ranking. The medium wasn't confined to the racecourse, and the "best" runner wasn't always defined as the fastest runner. Professional surfer Phil Edwards had once famously quipped, "The best surfer is the one having the most fun." For some individuals, running went beyond racing. These folks seemed to extract an inordinate amount of joy from running, the pavement and trails becoming their canvas, and they painted a picture that was both colorful and deeply personal. In their wanderings, running was a creative expression that led them to unexpected places and startling encounters. While racing is a sport, their running was more akin to art.

People such as Rickey Gates, Katie Visco, and Charlie Engle

were the beatnik poets of running, and they did things like running self-supported across America, running from the sea to the summit of the highest peak on all continents, and running across Australia. Their pursuits were adventurous and inventive, like a different form of running. I'd embarked on a few of these types of expeditions and loved the freedom and originality such endeavors afforded. You got to make up a starting line and the finish line, plus everything in between.

Sure, the Western States 100 was more structured than this, but in the early days it was still something of a mystical voyage. Though the passage of time had changed things and the sport had grown up, ultrarunning remained the funky uncle of traditional road racing. Still, I had to be honest, it was now more about racing and competition than ever before.

However, for me the upcoming Western States would be less about competition than simply finishing at all. I needed to take it seriously, not because I wanted to but because I had to. Gone were the carefree days when I could rely on the strength of youth to get me through. That era was over forever. Stakes were higher now. Getting into the race was nearly impossible, and my last go of it was disastrous, resulting in a DNF. The money was in the bank—at least as much as could be deposited in the short duration leading up to the race. Now it came down to not screwing things up; I couldn't afford to needlessly blow the budget by doing something crazy and uncalculated. This meant properly tapering. In younger years I'd carelessly stretch a 10-mile tapering run to 20 in the final days preceding a big race, but that was then. No longer did I have the deep pockets to cover such heedlessness.

Wisely, this time around I decided to obey the rules of taper and slow down in the week leading to race day. With five days remaining I departed on an easy 10K run, which would be the longest outing before the gun went off in Squaw Valley on Saturday morning at 5 a.m.

That final run felt decent. The muscles were loose, the body nimble, and the heart rate normal, no indications of tiredness. Maybe this would be my year. In twelve previous goes at Western States I'd never had a particularly strong performance. Each time seemed flawed, not the race I was hoping for. Perhaps my lighter than normal training would prove beneficial this time around. I hadn't run a single 100-miler all year; in prior years I could run two or three a month. Maybe this was the right formula. Race less = race better. Emil Zátopek—often referred to as the hardest-training runner of all time—never rested before a big race. Then in 1950 he suffered acute food poisoning and was bedridden for two weeks prior to the European Games. Despite this loss of training he went on to win the 5,000 meter and remarkably also the 10,000, comically lapping the entire field during this race and logging the second-fastest time ever recorded. From that point forward, Zátopek made it a priority to rest before races. Yes, maybe this would be my year.

The unhurried 10K taper run went by smoothly. When I returned, Julie was at the house.

"How was your run?" she asked.

"Good. I got a bit swept up in it."

"Those are the best."

"That's perceptive coming from a nonrunner."

"Well, I've known one for a while," she said with a wink.

I poured some water.

"You gonna come watch that companion of yours suffer this weekend?"

"Nope."

"Really?"

"Not gonna happen. That chapter's for you and your son to write."

I took a sip from the glass. "Then I'm in trouble."

She scorned, "You need to give the kid more credit."

"Credit? We haven't even discussed a plan."

"Just tell him what you want," she said assuredly. "He's like you, he listens."

"I'm worried."

"You shouldn't be. You're in able hands."

Kids grow up slowly, then all at once. Sometimes it's hard for a parent to keep pace. One day you're changing their diapers and the next they're teaching you how to use Instagram. It's not easy accepting that this once helpless youngster is now capable of helping you. The transition takes place for many years, then overnight.

There were other parents that helicoptered over their kids, swooping in at the first sign of distress. I was more like a satellite parent, orbiting high above. Maybe there was a happy medium somewhere within earth's gravity field. Perhaps I was too distant, especially in the formative years when they needed me most.

I remember when the kids were young and dependent on us for everything and wishing that they were a little less so. Balancing it all could be overwhelming, and at moments I found it challenging to remain patient and nurturing. I suspect most parents feel this way at points, but I remember secretly longing for the day when the kids were no longer at the house so I could focus more on training.

Now I look back on myself with disgust. Not only do I feel guilty for harboring such thoughts, once the kids were actually gone from the house I missed them terribly. How, even in a weakened state, I could ever wish for their absence was sickening. A father's relationship with his son is paramount in the development of that youngster. Did I do my job? You only get one chance at screwing up your kids. Had I committed this mortal sin?

Come Saturday morning I would have my answer.

14

TO CUT IS TO HEAL

Suffering brings salvation.

Three days before Western States I finally connected with Nicholas about our crewing plan. I'd written out instructions and thought we'd go over things item by item, so I set out a couple of pens, assuming he'd want to jot down some notes in the margins. That, at least, was my vision.

Nicholas apparently saw things differently. "Those your instructions?" he asked.

"Yes, I wrote everything out. I thought that would make it easier for us to go over."

"Okay, Pops. Thanks." He started walking out the door.

"But don't you want to go over it?"

"Whaddya mean?"

"I mean, don't you want to go over the instructions? I'm leaving later today."

He looked at me incongruously. "You wrote it all out. What's to go over?"

"Nicholas, do you even know where Michigan Bluff is?"

"I remember Michigan Bluff."

"Nicholas, you stepped out of your grandparents' RV and someone told you this place is called Michigan Bluff. They handed you a squirt bottle and a sponge and said to spray your dad off when he comes running in because he'll be hot. Then you got back in the RV and mysteriously arrived at the next aid station."

"Dad"—he patted me on the shoulder—"I'm not eleven years old anymore."

A moment of silence ensued. "But don't you want to go over things?"

"Dad." He glanced down at his phone. "I've got to go to work." He opened the door. "See you in Michigan Bluff." The instructions and the pens remained neatly lined up on the coffee table, untouched.

Julie came in on her way to work.

"You all set?" she asked.

"I'm concerned about Nicholas."

"Don't be. You've got your own job to get done. Good luck."

"That's it?"

"And return with your shield, or on it," she said with a smirk.

Later that morning I traveled to Squaw Valley—starting arena of the race—with my longtime friend Kim Gaylord. Kim would be pacing me at the race and we had an opportunity to catch up during the four-hour drive from the Bay Area to Squaw Valley (yes, I was bumming a ride with her).

While I knew a lot of people in the running community, Kim knew what was going on in the running community (i.e., the juicy stuff). We both spent a lot of time on the trails, but Kim had all the dirt.

"Did you hear about Stuart and Maxine?"

"No, but do tell . . ."

"And what about Valarie and François?"

"Seriously?"

I talked about the kids and the family; Kim talked about the things going on in the bushes behind the aid station. Apparently ultramarathoning could be something of a steamy cauldron. Long-distance running without the loneliness, birds and bees running amok. It all sounded delightfully scandalous, but better sport for a younger athlete's generation. My seeds had been sowed long ago.

Still, I didn't interrupt her.

Eventually I was brought up to speed on who was shagging whom, and the conversation turned to the next logical topic of primordial human urges, food. We were passing through the rural township of Auburn—finishing point of the Western States 100-Mile Endurance Run—and I knew of a local eatery and country market that was a favorite with the locals called Ikeda's.

For a roadside farm stand, Ikeda's had conjured up quite a tantalizing menu of items, such as Dungeness Crab Sandwiches with Kennebec Potato Fries and Gourmet Wagyu Burgers with Rosemary Sweet Potato Wedges, along with a plethora of freshly baked fruit pies, by the slice or by the pie. With such a gluttonous menu, Ikeda's is admittedly a more suitable postultra eatery; still, we ordered some Truffle Parmesan Fries because, well, they're amazing. We sat outside under roof-mounted misters eating epicurean fries and talking about how this place hadn't really changed much since the early days, Ikeda's nor the town of Auburn. Kim and I had been coming here for more than twenty years and Auburn remained a slower-paced enclave of older pickup trucks and ranchers. Though once a year the city gets inundated with a lively group of all-night dwellers hunting around at 2:00 a.m. for food and resupplies and speaking a foreign language (even if that language is English). The usually sleepy city snaps to life as the high school stadium fills with thousands of cheering revelers from around the world, all there to welcome a group of zombielike humanoids clad in brightly colored running apparel, covered head to toe in dirt and scratch-marks from thorns. This weekend was that one time of year it all happened.

Abounding with myth and legend, Auburn was first settled during the California Gold Rush in 1849. "Go west, young man" was the rallying cry. "There's gold in them thar hills!" It was a barbarous and untamed land, with wild-eyed prospectors looking to strike it rich. Some did, most did not. Others chose less savory paths to fortune, guys such as Rattlesnake Dick and Cy Skinner, who looted and robbed their way to riches before ultimately meeting their fate at the receiving end of a revolver. Rumors still circulated about buried treasures in the nearby hills, gold stashed away but never claimed, just waiting to be discovered.

Then there was Black Bart, the notorious poetic bandit that struck from the outskirts of Auburn, often leaving emotive epistles at the scene of his crimes.

I've labored long and hard for bread,
For honor, and for riches,
But on my corns too long you've tread,
You fine-haired sons of bitches.

You could just imagine the bullets flying. "Take that, you sons of bitches!"

Nowadays Auburn is known as the "Endurance Capital of the World," essentially trading one form of dusty raucousness for another. Trails are now trodden with Vibram running shoes rather than leather boots, but other than the modernization of footwear, the spirit of the Wild West still looms large in Auburn.

The drive to Squaw Valley from Auburn can take nearly two hours and travels along an expansive tree-lined roadway through the foothills of the Sierra Nevada and eventually up and over Donner Pass. For a runner, knowing that you'd soon be attempting that same passageway on foot is psychologically traumatizing. You suddenly feel very small. Arriving at Squaw Valley is equally intimidating. A colossal Olympic torch marks the entranceway, flames erupting skyward, the towering cornices of the surrounding peaks framing the skyline, the railcar-size gondola looking puny and diminutive compared to the gigantic mountaintops, like a tiny dung beetle scurrying up the leg of an enormous elephant.

The first thing I wanted to do, had to do, when we arrived was run up to the top of that gondola. Otherwise the intimidation festers, sleep becomes troubled and restless. Better to slay the dragon now rather than waiting till race day, to take the edge off the anxiety if nothing else. Never mind it was merely the dragon's toenail that was lopped off during that amble to the summit, a few short miles of the 100-mile footrace. Better not to think of it in those terms.

When I got back I checked the #SeeYouInSquaw hashtag. It was trending; race buzz was building. People were arriving from across the globe, a sleepy summertime Squaw Valley was starting to rub its eyes, yawn, and sit up. The dragon had come to life.

My parents pulled in later that afternoon. Wait, I never mentioned my parents were coming? My oversight. Though such an omission hardly seems noteworthy, forgone conclusions rarely warrant much preamble. More noteworthy would be if they *weren't* coming to Western States. This was a family affair, after all, like a Sunday picnic or a holiday gathering. Family was an essential ingredient in the recipe, and my parents had been to every single Western States race I'd done.

"This setting looks familiar," I said when I saw my mom.

She turned to hug me, but my dad swooned in from the side and we abruptly found ourselves in a socially awkward three-way embrace. People stared. I felt self-conscious for all of about thirteen milliseconds. Who cared what other people thought? Such was the gift of age.

Then my dad boomed, "Ultramarathon Man!" and suddenly I felt *very* self-conscious.

"Dad, pleeease," I whispered. He was like a child sometimes, excitable and prone to spontaneous outbursts.

We were standing in the grand foyer of the Inn at Squaw Creek, our customary meeting place. They'd driven up from Southern California in their RV, camping along the way, like two hippies, except they were old enough to remember when there was no Squaw Valley Ski Resort (it was completed in 1949). As tradition had it, we sat on the outside deck and had a glass of wine watching the sunset over the mountaintops, reminiscing. Twenty-four years prior we sat at this exact same table. The view looked different now, sweeter, like I'd earned it. My parents kissed at sunset, and when we walked out to their RV they were hand in hand.

It was my dad that first introduced me to long-distance running when he ran the inaugural LA Marathon back in 1986. He told my mom and me to meet him at the 20-mile mark along the course because at that point of the race he would hit "the wall," a metaphorical blockade that would require great willpower and resolve for him to break through. When we saw him, he instructed us to tell him, "You look good. Keep going!"

When he ran up to us at mile 20 my mom looked at him and said, "You look terrible. Why don't you stop?"

"Mom!" I couldn't believe she'd just said that.

"Well, he does. I'm just telling the truth."

Honestly, that's what I thought, too. The guy was a wreck. He truly did look terrible and probably should have stopped. But kudos to the man, he kept going. And when we finally located him in the medic tent at the finish line, lying supine on a cot wrapped in a Mylar safety blanket, snot streaming down his nose, I resolved to myself to *never* do anything like this myself. Ever.

The entire drive back to our house he was moaning and groaning, as though he'd picked a fight with Bruce Lee and very much lost. His body appeared puffy and swollen, like it'd just been dealt a series of sharp and powerful blows. Even his face was gaunt and disfigured, each eye encircled in a purplish-black ring. Periodically he would jerk spasmodically as a fresh wave of cramps rippled through his torso. He lay strewn across the entire backseat of the car and he squirmed impulsively like a gaffed fish, his agonizing yelps and contortions heightening with each mortifying new round of muscle contractions. *Never*, I thought. *Ever.*

But now here I was, about to embark upon a race that made the LA Marathon look tame in comparison. Something in those pained whimpers of his must have resonated with me. Or maybe it had more to do with the aftermath. Dad seemed a changed man after that marathon, more at ease with himself, less tormented. Suffering brings salvation, it's been said. My father had found his peace.

My parents met not far from here, in Lake Tahoe. My dad proposed to my mom within twenty minutes of meeting. She accepted and they wed. She was just twenty-one when she had me. My brother, Kraig, came along thirteen months later. Pary, our sister, arrived a couple of years after that. We grew up in Southern California. Kraig and his family still live in SoCal, near my parents. My sister lives in heaven.

No worse pain can be inflicted upon a human being than the loss of a child. I think that's why he did it, my dad, ran that marathon. Something beyond going to church and receiving communion had to lessen the hurt. He bared his soul at that marathon, opened up and welcomed the pain as never before. Yes, he looked terrible, and yes, his feet were bloody and pulverized, and yes, his muscles screamed in agony, but with each forward step the hurt somehow loosened its grip. He did it for her, to honor her rather than mourn her, that's why. And while he lay in that medic tent, entirely spent with nothing more to give, Pary was looking down upon him with pride.

15

BACK TO THE START

The sweetest fruit is well ripened.

The next morning was the prerace meeting. If you've ever been to such an affair you know they can be rather stiff occasions, like the early stages of a wedding. Emotions naturally run high and interactions seem forced and obligatory, smiles strained, almost compulsory. Nobody appears at ease, and exacerbating this particular situation in Squaw Valley was the fact that the assembly hall where the gathering was held was cramped and overflowing, with poor ventilation and lousy acoustics. The room became stuffy very quickly and it was difficult to decipher what the announcers were saying. People squirmed awkwardly, uncomfortable and confined. Here was a cadre of the world's most elite long-distance runners awkwardly constrained, unable to move or stretch out, trying to pay attention to inaudible mumbling over scratchy loudspeakers. The pent-up energy inside that building felt like it could blow the roof off.

One dynamic I noticed almost immediately was how international the field had become since I'd last run the race nine years ago. Because the lottery system showed no favorites, and entries from abroad numbered in the thousands, this shift to a greater percentage of overseas runners was a natural outcome. I wondered how many people in that prerace meeting had any comprehension of what was being said. The other element that had changed over time was the number of first-time participants. Here again, this was a mathematical function of the lottery system. With such a high ratio of the entries coming from first-timers the number of inaugural runners accepted into the race was naturally going to be higher. There were fewer familiar faces in the crowd than in the past because a higher proportion of the attendees were there for the first time.

In the 1990s, when I first started running Western States, half the field was from Northern California. It felt more like a local race back then, just on a grander scale. In an effort to grow international participation, the race started setting aside twenty-five spots for overseas runners to attract more foreign participants. It worked, and Western States became the most international of long-distance trail races. Then, in 2003, a race was launched in Europe called Ultra-Trail du Mont-Blanc (UTMB for short). The race starts and finishes in the mountain town of Chamonix, France, in the shadow of Mont Blanc, and passes through Italy and Switzerland. It took several years for the event to gain popularity, but once it did, UTMB absolutely exploded. While Western States was restricted to 369 runners, UTMB had no such limitations and suddenly there was an ultramarathon with not hundreds of participants but thousands of participants from around the globe.

And the sport itself took off in Europe and internationally. In

2019, France had the highest percentage of ultramarathon finishers globally (12.4 percent compared to USA's 12.1 percent).* More French finished ultramarathons in 2019 than Americans, yet France has only a fifth of the population. Globally, nearly 90 percent of ultramarathoners now are from countries other than the United States.

As I sat in that airless auditorium at the prerace meeting, it was clear that most people had come from someplace else. No longer were the majority of participants from Northern California, or even the United States. In many ways, the Western States had become the melting pot of ultramarathoning.

Although the prerace talk had been mostly muffled and undecipherable, one main point that was emphasized repeatedly was the heat. It was going to be hot tomorrow on race day, hot, hot, hot, and participants were warned to maintain proper hydration. In fact, if the forecasts were right, it could be the hottest race day on record. *Good*, I thought, *heat is my friend*. Besides, it was like a sauna in that auditorium, which was good acclimatization.

Looking around the room, I wondered what some of the other racers were thinking. This was familiar territory to me. Western States, Squaw Valley, and all the lore surrounding the race was now part of my DNA, interwoven into the very fabric of my soul. At the prerace meeting in 1994, my stomach filled with butterflies, I remember listening intently as the announcer described the surely impossible idea of running 100 miles through the mountains and arriving in Auburn the next day. Now that I'd done the race so many times, a bit of that wonderment was lost,

---

* Data provided by the International Association of Ultrarunners (IAU) and RunRepeat.

as could be expected with such familiarity. But I was curious as to what was going through the minds of the other racers. Did Western States affect them the way it had me the first time I had ventured to these hallowed ultrarunning grounds?

When the prerace meeting finally concluded and we runners and our crews began exiting the hall, I decided to ask someone in the audience their impression. A nice-looking gentleman, slender and tanned with dark, wavy hair, was walking next to me. "What did you think?" I asked him. "Is this your first time at Western States?"

He looked at me wide-eyed. "It tis fantastico," he said. "I never been to US before. Fantastico."

I smiled. "What's your name?"

"I am Maximo, from Italy." I figured as much. He was wearing La Sportiva shoes and a Diadora jersey.

In his eyes and demeanor I could sense his absolute captivation. It was new and unknown to him, scary and incomprehensible, yet wildly exciting. And I must admit, seeing his fascination enlivened me. He felt those same heightened emotions and pulsing adrenaline I had experienced twenty-four years earlier; I could relate to it all.

Then he asked me, "Will you please sign, ah?" He held out his racer's manual.

I smiled. "Sure. Do you have a pen?"

He patted his pockets and then turned to a youthful-looking woman standing next to him. "Do you have a pen, ah?"

I asked him, "Do you two know each other?"

"Yes, ah," he said. "This is Lucia."

"Hello," she said to me in a thick Italian accent.

"*Ciao bella.*"

They both stared in amazement. I'd been to Italy half a

dozen times, so I knew a few rudimentary phrases (emphasis on *rudimentary*).

Lucia didn't have a pen, either, so I suggested that perhaps we take a picture instead. "Do you have a phone?" I asked Maximo. "Let's take a photo."

"Yes, yes," he answered, pulling a phone out of his pocket and handing it to Lucia. He seemed nervous.

"Why don't we have my mom take it," I suggested. "That way all of us can be in the picture." I turned to my mom. "Would you mind taking our photo?"

"Yes, ah. Your mama."

They both seemed very happy that my parents were in attendance.

Maximo handed my mom his phone and stood next to me. Lucia was next to him. I suggested that perhaps Lucia should stand between us. "A rose between two thorns" was my lament. They both thought about my statement for a moment and then nodded and seemed to agree once the analogy codified. Lucia scooched between us.

My mom held up the phone and started taking pictures. It was too much for Dad to bear. "Fran," he said, "you've got to point the camera at their faces. You're taking pictures of their feet."

"Shhhh," she rebuffed.

He rolled his eyes. "And you've got to count down so they smile."

"Will you please!" she reprimanded him.

He sulked and moved back to watch. But the isolation was intolerable. "Okay, I'll count down," he chimed in. "Say cheese . . . Say pesto . . . Say Ferrari . . . Say see you in Auburn."

All the while my mom was randomly snapping photos. There was absolutely no synchronization between the two. He was spewing ridiculous catchphrases and she was repeating "Shhhh!" with her finger held steadily on the trigger. Yes, their marriage was made in heaven, but so are thunder and lightning.

Maximo and Lucia weren't exactly sure what to make of the proceedings. For me, it was family feud lite. You should see the two when they really get after each other. My mom will tolerate my dad only up to a point; beyond that, look out. She'll shovel it right back.

When she handed Maximo his phone he seemed relieved. He thanked her profusely, and then turned to me. "Your book, ah. I read in Italiano. I start running."

I thanked him. My books were gateways for some runners; the stories they told provided an introduction to ultrarunning and opened the mind to the limitless possibilities of the human body and spirit. It was gratifying to meet individuals such as Maximo and to see their enthusiasm, though I often wondered what it was like for those people to meet me. When you read about someone from afar, they tend to grow larger than life. I remember meeting the late Anthony Bourdain, and my first impression was that he looked tired, like a good, long nap was in order. Here was a guy I practically worshipped and the first thought to occupy my mind is that he should lie down and rest for a while. Suddenly he seemed very human, which oddly made me feel acutely aware of my own humanity and the supreme oneness of all humankind. Despite his immense talent, he wasn't a deity after all, he was someone just like you and me. If he stayed up late, he looked tired. And thus quite unexpectedly our encounter was infinitely more satisfying than had he actually been able to walk on water.

My reflection was interrupted by a commotion. Off in the distance, behind the auditorium, there was a bear digging through a Dumpster. His haunches were on the ground and his front paws were on the brim of the container, head buried deep within, sniffing around. Apparently a group of Japanese runners thought this an ideal photo opportunity. They moved in closer, with their phones held before them. *I'm not sure that's such a good idea*, I wanted to say.

It was too late. The bear pulled his head out of the bin and looked directly at the crowd. It was as though a powerful explosive had detonated. A percussive shock wave expulsed outward, and the entire group of onlookers recoiled in terror, arms flailing and feet frantically backpedaling. Once the initial terror rippled through the crowd a more clearheaded dispersal commenced. Some ran sideways, watching the creature over their shoulder as they made their escape; others just turned and bolted as fast as their feet would carry them. Remember, you don't have to outrun the bear, you just have to outrun the person next to you. The initial impact of the bear lifting its head had left several in the Japanese party on their butts, legs unable to retreat quickly enough. I saw one man being hauled backward by a friend, legs stiff and outstretched, heels dragging along the ground as his buddy attempted to tow him to safety, like a soldier rescuing a fallen comrade. There were screams of horror and deep gasps. Never in my life had I witnessed a crowd scatter so quickly.

The bear, meanwhile, seemed to find little amusement in our panic, nor did it sense any real concern and thus casually inserted its head back into the trash bin to continue rummaging around.

Given the choice between human flesh and yesterday's leftovers, most bears in the region would gladly choose the latter. If fact, if you were to set off a cherry bomb behind that bear it would most likely swat you away in annoyance and continue foraging for scraps of rubbish.

That's not to say that all bears in this area are harmless. Quite the contrary, especially if you were to startle one. In 2006 I came around a corner in the trail at about the 75-mile mark and saw a flash of brown and a huge dust cloud. A horse had gone lame in the path and was thrashing around, struggling to get back up. Unaware of what to do, I approached the dust cloud. No sooner had I reached the edge than a massive brown bear thrust its head from the dusty fog. That was no horse! It all happened in slow motion, like watching a hologram materialize. We were face-to-face, so close I could clearly discern the bloodshot sclera of its eyes. The response was immediate and severe: it went plunging headlong down the steep embankment, crashing through the dense thicket with the force of a runaway bulldozer while leaving a swirling cloud of dust spiraling skyward. The clear-cut path through the knotty undergrowth left in its wake looked as though a Volkswagen had been pushed over the cliff. The thing probably weighed as much as one, uprooting everything in its way. Breathless and shocked, I sprinted down the trail, madly trying to put distance between myself and that smoldering gouge in the earth. Only later did I fully process how lucky I was that the thing hadn't turned that same destructive force my way.

Wildlife encounters in the area aren't surprising. Remote and untamed, the Western States Trail was mostly still in its nascent state. Extending from Salt Lake City across the Great Basin and over the Sierra Nevada, the route was first traveled by the indigenous Paiute and Washoe communities. In the 1800s prospectors began using the pathway more regularly and foot traffic increased, though it was still quite treacherous and fraught with peril. When a survey party arrived in 1863, William Brewer noted, "It was a grand, but very rough trail, in fact, awful bad . . ." Not much has changed.

In Europe, the UTMB trail system is a popular destination,

with thousands of runners and hikers traversing it more or less continuously throughout the season. The Western States Trail sees far fewer. While there was once a thriving population of brown bears in the Alps, now there is only a handful in the whole region. There are plenty of bears left in the Sierra Nevada, as I've observed. The area remains a mostly wild and uncultivated place. And along with bears, there are other potentially dangerous creatures as well. I've seen many rattlesnakes along the trail, as have others. In 1994 a runner was tragically killed by a mountain lion while training along the Western States Trail. A memorial bench was built trailside in her honor, and we racers pass by it along the course. I've never seen a mountain lion myself, but they say you never do.

After the Dumpster-diving bear incident we decided perhaps it best to avoid large gatherings. Taking refuge in my parents' RV, there was really very little that needed to be accomplished today other than complete and total relaxation. I took to reading the morning paper, and my parents each buried themselves in their books. Their RV was stuffed with every imaginable sundry item. In the front cup holders there was some loose change, several golf tees, a small plastic magnifying glass, a few rubber bands, seashells, chewing gum, extra keys (for who knows what), a penlight, and an eyeglass repair kit. On the dashboard and in the side storage panels were maps, binoculars, guidebooks, headlamps, golf balls, a tennis ball, more seashells (bigger ones) and a couple of sand dollars, a multitool kit, flashlights and spare batteries, ballpoint pens, felt pens, Sharpies (of all varieties of colors), pencils, several halfway-finished crossword puzzles, a few dried flowers, a slide rule, and an electric razor. None of it was very well organized.

"Are those reading glasses?" I asked my dad.

"Yes, you're welcome to them."

They were on the floorboard in a cheap leopard-patterned eyeglass case of the sort they send you as an extra bonus when enrolling in Medicare or something. I took them out of the tacky case and put them on. They were actually quite comfortable.

"Where did you get these?" I questioned.

"I'm not sure. We keep lots of reading glasses around."

I looked at them closer. They were smudged and a bit scratched but vaguely familiar. "Dad," I shrieked, "those are my glasses!"

He looked at me quizzically. "Oh."

His easygoing attitude was off-putting. "You must have taken them from our house on one of your road trips."

"Sorry, son. Like I said, we keep lots of reading glasses around."

"Dad, those are two-hundred-dollar prescription glasses!"

"That explains it," he said, seemingly relieved. "I thought my eyes were getting worse every time I wore them."

"Did you not hear me? Those were two hundred bucks!"

"Have 'em back, that's fine."

"Jeez, thanks." They'd been missing for years.

I wiped the glasses clean with my shirtsleeve and went back to reading. There was no changing the man at this stage of the game, so why get upset?

After finishing the paper, I looked around the RV. It was more than a second home to them; it was their primary residence. The cabinets were stuffed with clothes of all varieties, from puffy down jackets to bathing suits. The bathroom had shave kits, lotions, shampoos, and conditioners, and the kitchen cupboards were filled with bowls and utensils from their house, many of which I used when living at home with them back in high school. The refrigerator was stocked with eggs, yogurt, hummus, and carrots, and many leftovers of unidentifiable composition that had been procured at roadside eateries somewhere along the drive from Southern California.

I opened one of the containers. "Mom, what is this?"

"Let me see."

I held it up for her.

"That's a veggie burrito from a little Mexican place we love in Salida."

"What's the green stuff?"

"Avocado. Have some."

"It looks like a smoothie."

"Things move around a bit when we drive. It's still good. Have some."

I hadn't had a burrito in years—since changing to a strict Paleo diet—but the thought of a burrito suddenly sounded very appealing, despite looking as though it'd gone through a blender.

"When did you get it?

"Yesterday?" She thought for a moment and then turned to Dad. "Honey, when did we get those?"

"What?"

"The leftovers."

"Which ones?"

"The burritos."

"We got them from that little Mexican place we love in Salida."

"No, *when?*"

"Hmm . . . sometime before we went to Sonora. Let's see, today's Friday. Maybe three days ago."

I closed the container and put it back in the refrigerator. "Do you have any fruit?"

"Sure," she said. "See those bags on the counter?"

There were three sealed paper bags on the counter. I inspected them. The color was slightly darker at the bottom, which was suspect. I lifted one of the paper bags, and the upper half broke free from the lower half, dozens of fruit flies emanating from the fracture.

"Oh, man! You've got a fruit fly infestation. When did you get this stuff?" I was standing there looking at her, holding the top half of the paper bag in one hand and swatting away fruit flies with the other.

"We got it at this little fruit stand we love near Manteca."

"How many days ago was that?"

She turned to my dad. "Honey, when did we get that fruit?"

"What fruit?"

"The fruit on the counter."

"We got it at that little fruit stand we love near Manteca."

"No, *when*?"

"Never mind, Mom," I interrupted. "It's fine."

"Just rinse them off," she said. "They're really sweet."

I looked at the gooey mess on the countertop with the circling fruit flies and could hardly decipher if they were peaches or apricots or some sort of disintegrated butternut squash. All of it had melded together.

"They're really sweet," she repeated.

Certain things didn't bother my mother, basically anything. There was really only one thing that could get her goat, and he was sitting not far away. And even so, he was more of an exasperation than any real source of angst.

Nothing bothered her. She had always trended in this carefree direction, my mom, but seven decades of life's tribulations had fully softened any remaining hardness. She was Buddha, this enlightened one, only more slender, and female, oh, and Greek. Other than that, they were about the same.

I looked at her, contentedly reading next to my father, and

reached for a piece of fruit. I washed it off under the faucet and took a bite. She was right; it was indeed sweet. Amazingly sweet. Though nowhere as sweet as the woman who had recommended it.

I've never known a person as universally loved as my mother. A retired middle school teacher, her job had been to educate youth at a time when hormones were raging, personalities frail, and authoritative figures looked on with disdain and mistrust. Even more challenging, she taught in a district where many of the students came from broken families and home life was quarrelsome and unstable. Still, these kids loved her. For years they would write her letters and correspond with her. I recall reading one such letter from a young man who had recently completed his cardiology internship. He explained that she was the only person in his adolescent life that treated him with any kindness or encouragement. I looked at her again; how lucky I was to have her in *my* life. Both of them. We had started this Western States journey twenty-four years earlier and still the journey continues. You cannot teach love and devotion and commitment; these things are acquired only through repeated application, only through unwavering loyalty to show up and immerse yourself again and again. And we'd been doing just that for the past two and a half decades. We knew each other now, intimately. We knew our strengths and our shortcomings, could almost predict how one would respond in any given situation. And that was accepted. What mattered most was that we continued the journey, always making an effort to be together and to keep moving onward along this adventure that is life. That is dedication, that is endurance, that is what living is all about.

When we eventually exited the homey sanctity of the Mother Ship, the lot was entirely vacant. It looked like a ballpark during the off-season. Nowhere to be found were the news broadcast vans with their satellite antennas erected, reporters dashing around interviewing race officials, or press conferences with notable athletes taking place against heavily logoed sponsor backdrops. Here was the most important ultramarathon in North America, arguably the world, and the starting line was deserted. We stood there, my parents and I, all alone looking up at the mountaintops in awe, wondering what tomorrow would bring, just as we had done twenty-four years ago. And that is what I loved about this race and about this sport and these people standing beside me. Sure, things had changed, but the magic remained.

16

LET'S GET THIS PARTY STARTED

An ultramarathon moves slowly very quickly.

The alarm went off at 3:00 a.m., though I wasn't asleep. They say two nights prior to an event is the most important. That's because nobody sleeps soundly that last night before a race. I'd proven this theory valid once again, tossing and turning the entire evening, sleep eluding me. Despite knowing the importance of sleep, the mind was unable to quiet. If you've ever run a race before, you've been here—the gun essentially goes off the moment your head hits the pillow; the start isn't until the next day, but the mental confrontation is already under way.

Groggily sipping coffee with my parents, we discussed the race plan. Instead of having them traipsing around throughout the Sierras to all the remote outposts along the early stages of the course, as in younger years, the arrangement this time was to have them meet me at Michigan Bluff—just beyond the midway point—which was the same location where Nicholas would rendezvous with us. My dad, of course, was unsatisfied with this plan. "How are we going to find Nicholas?" he questioned.

"I think it'll be easier for him to spot you. The Mother Ship is hardly inconspicuous."

"I'll call him."

"Remember, Pops, there's limited cell reception in that area."

"I'll call him beforehand."

"He communicates via text. Kids don't talk these days."

"I'll text him."

"Pops, I've seen your text messages. I doubt the world's leading cyber security experts could crack that code."

I had another sip of coffee. "If you're worried about him use Find My Friends."

"What's that?" my mom asked.

"Here," I said, "let me see your phones."

I set up Find My Friends on their phones and gave them back. "That dot is Nicholas." I pointed at the screen.

"Nicholas is that dot?" my mom asked.

"Well, he's not *actually* that dot, but that dot is where his phone is located, and his phone is always on him."

"It's not moving," my dad butted in.

"That's because he's probably asleep, Pops. He doesn't need to be to Michigan Bluff until this afternoon. I assure you that dot will move once he starts driving."

"I don't trust it."

"Okay Pops, here's what to do. When you get to Michigan Bluff, start collecting kindling from the roadside. Pile it all together and ignite it. Then take one of those beach blankets from the Mother Ship and regulate the release of smoke into the air. Use those transferrable skills from your Morse code days in the army."

"Very funny," he said.

"Seriously, smoke signals would probably be easier to interpret than your text messages."

My mom was nodding her head in agreement.

The jest was all in fun. I held up my coffee mug. "I love you guys, thanks for being here today." We toasted, and then my dad, forever the nostalgic, hailed, "Let's get this party started!"

I took the cue and hit the play button on my phone. And the music began. It was Zorba the Greek, of course. We rose to our feet, locked arms, and started the slow, circular dance: small sidestep, kick, big sidestep, kick, backward step, bow, small sidestep, kick, arms held high, spin . . . Faster and faster the tempo progressed, and quicker and quicker our movement hastened in lockstep, snapping our fingers and kicking and twirling with increasingly unrestrained gusto as the beat accelerated, our contortions and gesticulations becoming ever more zealous, each bend and kick further accentuated, heads twisting left, then right, sweat flying from the brow. It was 3:30 a.m. when the song came to a magnificent crescendo, whereupon we breathlessly hailed, "OPA!" The party had officially started.

On the way to the start it seemed evident that the weather forecasters had been accurate in their predictions. The dark, early morning air was motionless and occasionally we'd pass through these little invisible thermal cylinders of warmth. Here in the high country nighttime usually brought with it cooling, but not today. This afternoon, in the canyons at the lower elevations, it was sure to be toasty—actually, hellish.

I carried with me two water bottles this time, both insulated.

One would be used principally to store ice. Heat was my friend, but so was ice.

Walking toward the starting line we came to a point where only racers were allowed to pass. We said our good-byes and they wished me well. "See you in 55 miles," I said, and they casually nodded their heads in accord. The ease by which that statement rolled off my tongue and the nonchalance of their acceptance illustrated the ludicrousness of ultramarathoning. We were standing at the base of a ski resort looking up at these imposing granite spires in the predawn light and I was about to run 55 miles through the mountains on a rugged and rocky dirt trail and meet them at some remote juncture called Michigan Bluff, and we said so long as though they were dropping me off at the office.

Walking to the starting area, nerves on edge thinking about everything ahead today, I kept reminding myself to keep it simple. There were just two things I needed to accomplish today: make it to the starting line, and make it to the finish line. I stepped into the starting corral: one down, one to go.

Once inside, the energy was kinetic. Sparks discharged into the predawn darkness, the collective pulse of the crowd palpable. I exchanged fist bumps with several people and said hello to one of the top female contenders, Courtney Dauwalter. She nodded her head and raised her eyebrows in a genial way that said, *I'm gonna kick your ass*. And no doubt she would. She would likely beat 95 percent of the men's field. In fact, in 1994 Ann Trason beat 99 percent of the men's field, finishing runner-up only to the legendary Tim Twietmeyer, and by the slightest of margins. And in 1995 she did the exact same. At shorter distances men's race times are predominantly faster. At longer distances the gender pace gap narrows. In fact, at races beyond 195 miles female ultrarunners are faster than their male counterparts. That's the wonderfully egalitarian aspect of ultramarathoning: women are on equal footing. It wasn't uncommon for a woman to win an ultramarathon outright, the playing field being more level for both sexes when race distance and duration stretch out. No, best to leave your ego at home in this game, gentlemen.

With three minutes left before the start, the mood subtly shifted, a transmutation was under way, the pleasantries and chitchat fading and a more savage, beast-like energy emanating from the crowd, pupils dilating, hearts pounding stronger. There is some long-ago forgotten element of the human creature that yearns to run wild through the mountains, chasing and being chased, muscles engorged, survival anything but certain. Domesticated as we've become, such primordial urges still dwell within us, and here at the starting line of Western States this animalistic temperament was discernible. We were no longer accountants, teachers, and businessmen; orderly life was about to be abandoned. In a few short moments we would be stepping into the wild and becoming our untamed selves.

When the final countdown began, people started howling at the sky, while others bowed their heads in quiet prayer. Some made the sign of a cross on their chest; others pounded theirs: AROO! AROO! Many runners just stared narrow-eyed at the trail ahead, focused and determined. Western States is the oldest 100-mile trail race in the world; it's where it all began, the pinnacle, the genuine article. *Three! Two! One!* The starting gun blasted and the animals were set free from the cage, 369 warriors surging forward in pursuit of glory. In the stampede I was kicking and being kicked, elbowing and being elbowed, refracted beams of light flickering through the dust cloud of our moving feet. Like most, I'd fought hard to get here. And like most, there's no place I'd rather be.

To say the initial climb was like running up a ski slope

wouldn't be an exaggeration, because that is precisely how the race begins. From the base of the Squaw Valley Ski Resort the course heads straight up to the highest peak. A few of the front-runners took off dashing uphill, but they were the exceptions. Many runners shuffled forward at a reserved pace, while the majority of the field power hiked. The path was wide and could comfortably accommodate three abreast, and groups of runners broke off into distinct packs clumped together, each runner on the heels of the man in front, shoulder to shoulder in pairs of two or three across. Conversations were limited, which was a sign of an experienced group. Running up a ski slope in the early stages of a long race was not the time to expend energy talking about the weather or ball game scores.

When we reached the top of the 2-mile climb the course turned a corner, and there was another climb. I knew this was coming but it was still menacing seeing it all displayed before me, more piles of granite, more broad clearings cut between pine trees, more chairlift cables disappearing skyward. The pace was so menacingly slow, the heart rate so rapid, the breathing so rushed, yet we were scarcely 2 miles into the race. Some runners moved quickly, and it was tempting to lock stride with them, but was it sensible? Sure, you'd get up the mountain faster, but nothing in this sport comes without a trade-off. You could find this same runner slumped over on the trailside somewhere down the way.

Like everyone else, my goal today was to finish as quickly as possible, but at least by 10:59:59 a.m. tomorrow, the official cut-off time. *Do nothing to sacrifice that objective*, I thought. I looked at my watch. Forty-five minutes had eclipsed since the start and I'd hardly gotten anywhere. *Patience*, I told myself. *Patience*. Respect the distance, take each breath purposefully, make every step count, calculate the surges, watch your heart rate, stay hydrated, maintain your calories. Do all of these things continuously while remaining incorruptibly patient. An ultramarathon moves slowly very quickly.

Eventually Emigrant Pass came into view, the highest point on the course at nearly nine thousand feet above sea level. The pitch to ascend that final climb was so steep it required scurrying on all fours, the hands clinging to clumps of undergrowth to help steady the way and assist in hoisting one upward while avoiding slipping downward. There is no particular path to follow in this section of the racecourse, and runners spread out across the landscape taking different lines to the top, the hillside assaulted by an army of crabs ascending a mossy seawall.

Once over the precipice of that sharp incline the final grade to the summit was a steadier path, though one covered in residual snow from the winter ski season. The first rays of morning sun shone from across the eastern horizon, mirrors of reflective light dancing across the still blue waters of Lake Tahoe, wispy fingerlings of pinkish clouds stretching overhead, the icy white snow underfoot taking on a rosy hue.

At the very top of the climb the view was unobstructed in every direction. A small group of individuals had gathered at this vista to watch the runners pass by. Some cheered, some took pictures, some just watched in admiration. The mood was faintly spiritual, spellbound by the morning brilliance and the reverence for the journey still ahead. I stopped alongside some other runners to take in the sights. "Where do we go from here?" one of them softly asked.

I scanned the horizon. "You see that peak over there?" I said, motioning to an impossibly distant point that looked to be about 20 miles to the west of us. "Once we get there it's about 75 miles to the finish." It was the same thing another runner had said to me twenty-four years earlier standing at this very spot, and the runner I told this to today looked about as terrified as I had been back then. Of course, I was currently feeling some level of those same terrors myself. Rarely in an ultramarathon is one affronted with such an arresting panorama of things to come, the unfiltered magnitude of the endeavor abruptly thrust in your face with a brutal honesty that says, *Don't tempt me, little man.* To think you've got to run from where you're currently standing to that faint peak on the horizon, and then run much, much farther beyond that speck quiets even the most brazen. Western States is a point-to-point race, not a loop or out-and-back route, and standing at Emigrant Pass above the tree line the entire course is displayed before you with instantaneous, shocking abruptness. Perhaps that's why so few runners stop to look.

"Auburn ho," I said to the others, and headed off down the backside of the mountain.

The downslope was an opportunity to regain some of the time lost during the lengthy climb up the front side of the mountain, the gently slanted decline being very "runable." A long shadow was cast in the lee of the summit and it felt cooler and dry, though very still, dust from passing runners languidly hanging in the air, only to be parted by the next darting passerby. Bounding down the path I glanced briefly at my watch and took note that my pace was just under six-minute miles, which was ludicrous, except the entire field was flowing at this same speedy rate. The path was narrow, and if I were to slow I would most certainly be rear-ended, which was probably the same thing the runner in front of me was thinking. We were *all* going too fast, but there was no getting off this train.

Eventually the trail opened up and runners were able to pass or be passed. Ironically, very little of this took place. The entire pack seemed to collectively moderate. It was still possible to run sub six-minute miles, though suddenly no one seemed all that interested in doing so. It was herd mentality, ultramarathon style. Wildebeests indeed we were.

The next section of trail we approached was notoriously tricky. It scooted along the base of a snowcapped peak, streams of snow melt pouring across the pathway with varying degrees of intensity. Personally, I don't like having my feet drenched at such an early stage of a race, which could hasten the onset of blisters. Sometimes wet feet were inevitable, but here they were optional. Thus I made every attempt to tiptoe around the deeper pools, which sometimes meant slowing down to analyze the best rocks to step across. This lasted for quite a distance and it was definitely costing me time, but I considered the trade-off worthwhile. Others were just charging through the rivulets with utter abandon. Fools.

Reaching the end of the watery section I was pleased that only the very tips of my toes had gotten slightly moist; I'd successfully evaded complete immersion. Mission accomplished. It may have slowed me down, but my prudence in navigating the watery minefield would most certainly pay off later. *I am so wise*, I thought. And then we came around a corner in the trail only to find a waist-deep river of water flowing across the path with no possibility of dry passage in either direction. Who put this here? In my prior race experience I don't recall a tributary of such significance at this particular juncture, though it had been nine years and I guess things could change. *Who's the fool now*, I thought, and plunged in.

With soaking feet, I arrived at the first aid station, Lyon Ridge. It was a scant 10 miles from the starting line and it had taken 2 hours 13 minutes to cover that distance, right about where I wanted to be. I'd gone out rather conservatively and all seemed in order.

However, the aid station was anything but orderly. There was much more activity than I'd recalled in years past; then again my pace was slower, which put me more in the center of the pack.

Aid stations, especially in the early stages of a race, are hectic and rushed places. The mood is frenzied with lots of anxious runners scampering around trying to grab the necessary food and liquid, while the aid station staff does its best to provide for everyone. In the pouch of one of my water bottles were some energy gels. After watching people bumping into each other and stumbling over one another trying to get to the food section, I decided to avoid it. I'd just eat a gel from my personal stash. I tore off the top and started sucking on it.

Even though it was early in the morning, the sun was already warming the land. In the pouch of my other water bottle were some UV protective arm sleeves. It was time to put them on. But before I could do so someone slapped me on the back and thundered, "KARNO!" It was one of the volunteers. "Welcome to Lyon Ridge. What can I getcha?"

He was a nice enough guy and was probably happy to see a familiar face. Many of the fifteen hundred aid station volunteers along the course hail from the West Coast, so it was a tight community of individuals. I handed him my water bottles. "Maybe fill one with water and the other with ice." He grabbed both of them from me. "Roger that," he said, and dashed over to the aid station supply table.

Looking around the crowd, and judging from the apparel and equipment of the other runners, I would say that about half of them appeared to be from someplace else. There were few formal greetings going on by name, and people treated each other with the respectful distance of a stranger. I saw some of the runners pointing to the things they wanted on the aid station tables, probably not knowing the word to tell the volunteer. There were lots of polite smiles exchanged, the kind used to express gratitude when you can't communicate with someone in the same language.

"Here ya go, man," my helper said, jarring me from my appraisal of the crowd. He handed back my water bottles and I thanked him. "Just one thing before you go," he added. "Do you mind if we get a quick pic?"

The request sent me aback. I thought about it for a second and then said, "Of course. That's fine."

"Hey, Rick," he shouted to one of his buddies, "can you take a picture of Karno and me?!"

"Sure, man," Rick responded, walking over to us. He wiped off his hands on the front of his shirt and took the phone, then moved back a step or two to frame us. But unaware of our intentions, people kept walking in front of him. When they eventually noticed that we were trying to take a picture they froze in that awkward position of someone waiting for someone else to finish something before going about their intended action.

Once they paused, Rick held up the phone. "Ahh," he stammered, "the screen's locked."

He brought back the phone to my helper, who unlocked it. In the meantime people had resumed going about their business. Rick took a couple paces back to frame us once more, and the cycle of momentarily freezing in place began anew. Those few seconds of suspended activity were unbearable.

"Got it!" Rick finally concluded, lowering the phone and walking back over to us. "I took a couple," he said. Then he asked, "Do you mind if I get one, too?"

This was turning into a production.

"Hey you guys," I said, "why don't we step over to the side so people can get what they need."

We moved behind one of the aid station tables to get out of the way. It ended up they both wanted to be in the picture, so I suggested we snap a selfie. This, apparently, was not a skill set the owner of the phone possessed, I concluded, after numerous botched attempts. Now people were starting to look at us like, *Who is this guy? Wait—I think I may recognize him. Should I get a picture, too?*

"Why don't I take the photo," I said hastily. "My kids have trained me well."

I set one of my water bottles on the table and took the phone. Holding out my arm I instructed, "Say CHEEEEEEEESE." I kept my finger on the trigger for a good long time to make sure we got a number of them.

"That oughta do it," I said, handing his phone back. They thanked me and we clanked fists. Then I set back out.

Once on the trail I fell into lockstep with another runner. He was from Seattle and it turns out we had raced together at a couple of other events. This was his first time at Western States. He said what he feared most was the upcoming heat later in the day. "Those canyons," he said, "I hear they're terrible." I concurred. With temperatures as warm as they already were, the canyon section of the course was sure to be sweltering. Though, honestly, the heat was already surprisingly intense, if you ask me.

It was nice having some company and we spent a mile or so running along pleasantly together. Things were all right; everything seemed to be going as planned, the heat perhaps a bit more than I'd projected, but manageable for the moment. Eventually he said, "I've got to find a bush about some business. I'll see you up the road." I wished him luck with both endeavors—the race and the business in the bushes—and continued motoring along solo.

Reflecting on our talk about the heat prompted me to take a moment to slow down and put on my cooling arm sleeves. Now was the time to be proactive in my approach; these early precautions would yield big returns. Then a wave of panic washed over me, like that sinking feeling you get when you can't find your wallet. "Shit!" I blurted. It would be impossible to put on those cooling arm sleeves, impossible because they were stashed in the pouch of my water bottle, the water bottle I'd left sitting on that table back at the aid station when I took our picture.

I'd come too far to turn around, so I kept going. I had a spare bottle at the Red Star Ridge aid station, though that was still a good distance away. And the trail between here and there was notoriously exposed, the scorching red earth absorbing the sun's sizzling radiation, magnifying the heat. Moments ago all was so hopeful; now it was a little less so. The Western States guidelines warn, "Carry ample fluids and be prepared, for the high mountains and deep canyons, although beautiful, are relentless in their challenge and unforgiving to the ill-prepared." I was now among the ranks of the ill-prepared. A barren, murderously hot section of trail lay ahead and I had limited UV protection, and the lone water bottle I did possess was just slightly filled with some melting ice.

This was my new reality. Suck it up, buttercup.

17

LOOSE LUG NUTS

If things are going well during an
ultramarathon, give it a moment.

All along I've been claiming that this was my thirteenth go at running Western States, but that's not entirely true. In 2002 Tim Twietmeyer and I set out to run the Western States trail in the middle of winter. That's not entirely true, either. Conditions were not so good that winter (i.e., extreme avalanche danger), so we postponed until the following year. But it was the same scenario that next winter: deep and potentially deadly snow accumulations. Finally, in January 2004 the inaugural winter Western States Endurance Run finally got under way. The party had expanded to include Bill Finkbeiner and Jim Northey, two experienced ultramarathoners and backcountry mountaineers.

Because snow covered the course we wore snowshoes, which resemble oversize tennis rackets that are strapped to each foot to prevent an individual from sinking into the powdery abyss.

The going was slow, though without snowshoes it would have been impossible. We wore heavy packs filled with food, hydration, extra clothing, and safety gear. Those extra supplies added more weight to each footfall, so even with snowshoes we'd periodically break through the surface and find an ankle or shin swallowed in frozen white quicksand.

A navigational beacon we aimed for was a landmark along the Western States course called Cougar Rock, which is about 13 miles from the starting point. After nearly half a day of back-country plodding through the snow we eventually came upon it. Little more than a conical nub, it made for a nice prop to sit atop and stretch our legs.

Today, in the summertime, Cougar Rock is an imposing lava rock monolith the height of a three-story building. Unburied in snow, it is one of the most iconic sights along the Western States trail and requires runners to scramble up the jagged rock face on all fours to ascend the cathedral-like uprising.

When my fingers made initial contact with the metamorphic surface I recoiled instinctively, the way one does when touching something that's been in the microwave too long. The sun had scarcely cracked the horizon and already the earth's surface was on fire. Part of the heating, I suspected, was due to the subtropical moisture that had streamed in from the south overnight. This was an atypical weather pattern, though not one unheard of. The usual atmospheric airflow over Northern California and the Sierra Nevada during this time of year is from the west or northwest, but periodically this current is altered, and warmer air from the Gulf of Mexico moves into the area. Along with this warmer air, solar radiation intensifies and humidity increases.

For a runner, humidity complicates matters. According to the army's heat index scale, if the air temperature is 102 degrees

Fahrenheit and the humidity is 55 percent, the effective temperature is 130 degrees, which is classified as extremely dangerous,
with "a high likelihood of heat disorders during prolonged exposure or strenuous exercise." And an ultramarathon is fundamen-
tally prolonged exposure to high temperatures during strenuous
exercise.

The reason humidity is so vexing is because of the way it
disrupts the body's normal mechanism of evaporative cooling.
Having no fur, humans are uniquely engineered to dissipate
heat through the skin, thereby keeping us cool. In fact, of all
mammals we are about the best at regulating body temperature.
This is why a human is capable of chasing down an animal, or
why Gordy Ainsleigh—founding father of the Western States
100-Mile Endurance Run—was able to outrun a horse.

But as humidity creeps upward, this dynamic starts breaking
down. The body's perspiration is less effectively evaporated because the air holds a similar amount of moisture to that on the
skin. And if you have no evaporation you have no cooling. Already I could feel the flesh on the back of my neck reddening, the
piercing rays of the morning sun cooking my pink derma the way
a chef uses a blowtorch to sear ahi. There wasn't much that could
be done to stay cool at this point. What little water remaining in
my bottle was warm and foul-tasting, the plastic container warp-
ing in the grip of my hand. When I tried to suck out that last drop
of liquid the mouthpiece was hot against my lips and the contents
tasted synthetic.

We followed an exposed ridgeline with the sun at our backs,
the rocky red soil underfoot perceptibly hot through the treads
of my shoes. The solar emissions seemed to zero in on this particular stretch of trail with sharpened intensity, and there was no
place to escape it, nowhere to lessen the ultraviolet assault. Off

to either side of the path were deep, yawning canyons blanketed in a low-lying vaporous mist that clung drowsily to the treetops. The air was thick and improbably stifling for this early hour of the day, so still and lingering. Already I could feel my body weakening, the once abundant pool of energy seeping away. A runner came up from behind and I pushed the pace, but it felt abnormally strenuous, my heart racing and my breathing deep and labored. I could feel the fabric of my sweat-drenched shirt clinging to my body, the technical material that was so proficient at wicking moisture away from the skin rendered useless and overpowered by the humidity. The army's heat index scale was proving its point.

The runner that had come up behind me hung tightly on my heels; the noisy sloshing of icy liquid in their hydration pack disrupting the reflective sereneness of the natural surroundings like a bartender mixing a martini in a crowded nightclub. I could almost feel a hot breath on my neck, two oversize exhaust nostrils blowing a fiery emission on my backside. Why they felt the urge to follow so closely was baffling—there were 100 miles of open track—but this individual apparently deemed it necessary that we should nearly trip over each other. Frankly, it was obnoxious, but no matter how much I altered the pace they tracked on me like a heat-sensing drone. On several occasions I said, "If you'd like to pass just say the word." No response, nothing. If I slowed, they slowed. If I accelerated, they accelerated. Obnoxious, but despite my best efforts I couldn't drop them, and I was laboring unnecessarily attempting to do so. Finally, after a mile or so of failed evasive efforts I pulled over to the trailside and let them pass. He held up a hand while moving by in a cursory, almost derogatory manner. It was a *he*, and from what I could tell he was a jerk.

I tried not to typecast, but judging from his Salomon trail shoes, RaidLight pack, and spectacularly tremendous nostrils, I had my suspicions about his origins. Such arrogance. I didn't like the guy.

Though enough about him, what was up with *my* deplorable state? It was too early in the race to be depleted in such a significant way; something wasn't clicking. Sure it was hot, but c'mon, this is a guy who's won the Four Deserts Challenge, a guy who's completed the Badwater Ultramarathon—a 135-mile foot crossing of Death Valley—on ten separate occasions. It might be hot, but it wasn't shoe-melting Badwater hot. Still, there was no denying my struggle today, no hiding my early fatigue.

The truth is, pedigree and prior accomplishments don't mean squat out here. Ain't no resting on your laurels; this is a sport that eats its contestants. Western States would chew me up and spit me out with spectacular indifference. Take your hubris and pile of buckles and shove them where the sun don't shine, Badwater boy. There are 85 miles in front of you, and you're gonna have to earn every damn one of them if you want another buckle for that fancy collection of yers. So better pick it up, *ultramarathon man* . . . I chugged on.

An interesting surprise awaited not far down the trail. It was my Salomon/RaidLight friend with the impressive nostrils. He wasn't doing so well, moving very slowly. I came up close behind him. Now he was feeling *my* hot breath on his neck. It was beautiful. He held up his hand in that same haughty manner as he stepped aside for me to pass.

Maybe it isn't arrogance after all, I thought. Maybe it was just his thing, this dismissive hand gesture.

"You all right?" I asked while shuffling past.

"It's not so good," he responded.

Honestly, I wasn't so good, either. But sometimes seeing someone worse off than you can lift your spirits.

"I'm sorry." We ran together for several paces. "Where you from?" I asked.

"I come from France."

Aha! I knew it. The shoes, the pack, that extraordinary schnoz, all dead giveaways.

"What is your name?" I questioned.

"Fabrice," he answered, pronouncing his name in three distinct syllables: *Fab-REE-cee* (hey, he's French).

"Zan wat tis yur name?" he asked.

"Dean," emphasizing the pronunciation as one tight syllable.

We shuffled along the trail together not saying much. It was an illuminating nonversation, as these footsteps with a complete stranger during an ultramarathon can sometimes be. There is a universal language that doesn't need words, an unspoken communication that transcends verbal expressions, and this connection is strangely affecting and profoundly human. It is during these nonverbal exchanges that real truths are revealed. And what it told me is that I didn't dislike Fabrice. In fact, I kinda liked the guy; he wasn't a jerk. As he tottered along in my slipstream, I could sense his weariness. His breathing was rapid and shallow, and he emitted a faint murmur at the conclusion of each strained exhalation. During the traverse of a few hundred feet of trail together my fondness for him turned to sorrow. He was clearly an experienced athlete, and looked to be in top physical condition. But he was crumbling, and it was still early in the battle. Perhaps my empathy was owing to the fact that I wasn't that much better off. I could step into his shoes and would probably feel only slightly more dismantled. Still, I had traveled from the San Francisco Bay Area to get here; he had come from France. It had to hurt.

"I cannot go," Fabrice said, slowing to a walk.

I slowed to a walk, too. "I will walk with you."

He took a deep breath. "No, you go. I cannot."

There was a subtle conviction in his tone that suggested I best be moving along.

"It was nice running with you." I turned to look him in his eyes. "I'll see you down the trail."

And his look signified he grasped my sincerity in that statement. I wasn't sure if he would keep going, but I really hoped that he would. In Buddhism there is a state of consciousness called metta, in which the soul of one identifies wholly with the suffering of another. At that moment along the trail I did not care to beat Fabrice in a long-distance footrace, I just wanted his suffering to end.

Arriving at the Red Star Ridge aid station, the wheels of my bus hadn't completely come off, but a couple of the lug nuts were certainly loose. How a mere 16 miles had so broken me was worrisome, though I tried not to think of that and focused instead on taking curative measures. Some parts of me needed reassembling.

Like the previous aid station, Red Star Ridge was a busy place. People rushed around as though an air raid siren had been sounded, frantic, almost hysterical. Runners kept constantly pouring in, overwhelming the system of orderly arrival and dispersal. The food tables were absolutely assaulted, with dusty, sweaty foragers seeking an ideal morsel to cram down their throat. It has been said that we are creatures raised by God above the animals, but just look at any aid station's food table during an ultramarathon and you may think otherwise. I went in with both hands. Savages we are.

After food, I sought liquid. My level of thirst was somewhere between keen and I'd drink camel piss. I needed water terribly, which was problematic because a line had formed to get to the jugs. I tried to be patient, feeling light-headed and faint. Scrunched in my pocket I carried this newfangled HydraPak reusable race cup; I pulled it out in anticipation. Finally my turn came and I held the cup under the spigot. It filled almost immediately and I gulped down the liquid. But it barely wet my mouth. The cup was advertised as holding seven ounces, but that must be if every molecule of H_2O is perfectly aligned inside. And because the cup is collapsible and highly pliable, when squeezed the sides compress inward, sending half the contents spilling over the edges. I stuck it back under the spigot and repeated the process, but my thirst was hardly quenched even after this second round. Now runners behind me were growing impatient; they didn't seem to care much about my efforts toward conservation using this eco-friendly thimble. Dejected, I stepped aside.

Thankfully an aid station volunteer tapped me on the shoulder. "Here is your bag." She handed me my drop bag. "Can I get you anything?"

I opened the bag and pulled out my spare bottle. "Could you fill my water bottles?"

She gladly agreed and asked me if I wanted to sit down. There was a hodgepodge row of various-sized folding chairs set up to the side of the aid station. Many of them were occupied, though a few were empty.

"I'd like to sit down," I told her, "but I'd better not." She found that comment quite amusing, then off she went to fill the bottles.

Volunteers are the lifeblood of this sport. So much of the heavy lifting is done by people who willingly give their time and energy to set up, staff, and dismantle aid stations, oftentimes in remote, mosquito-infested, cold, and isolated locations, at every imaginable hour, in every conceivable weather condition. And all this happens before the first runner even arrives. Once the floodgates open, things really get hoppin'. There is something about the spirit of volunteerism that brings an aid station to life, some mystical pixie dust that elevates the experience beyond that of a mere feeding trough. I've spent plenty of hours constructing and staffing aid stations and I will tell you that once the athletes begin arriving, positrons start to fly, the energy gets ratcheted up, the place instantly comes to life, people serving and being served, warriors entering and exiting, sharing laughter and tears, for this is a place of high drama, a place where dreams are realized and dreams are dashed. Theater is the imitation of life; an ultramarathon aid station is the live performance, real and raw.

By the time the volunteer returned with my bottles the seating area had completely filled. "Is there anything else I can get you?" she asked.

"Not sure. I'm having a rough go of it today. I don't know what's wrong."

She inspected me. "Let's use Allen Lim's HALT assessment."

"HALT?"

"Yeah. Are you hungry? Angry? Lonely? Or tired?"

I scratched my head. "All of the above."

"Here, this always helps," she said. "We're giving out hugs today."

She gave me an embrace, which I thought was very courageous because I probably smelled like a mule. It felt remarkably uplifting, better than any energy gel, even the ones with caffeine.

"Wow," I remarked, "that tops a Wellesley kiss."

She had absolutely no idea what I was talking about, but I

smiled assuredly, as though topping a Wellesley kiss was something of merit, which indeed it was, and she seemed satisfied with that.*

"Thanks for all you're doing today," I said with a nod, and turned and walked away.

Exiting the aid station someone called out, "Three eighty-eight."

And then another voice hailed, "Karnazes out."

A tick was marked next to my name on a race clipboard, and thus I had officially cleared the Red Star Ridge aid station.

I turned back one last time to take in the scene. Fabrice, my French trail buddy, still had not arrived.

---

* Wellesley College, a private women's institution, is at the midpoint of the Boston Marathon. There is a tradition that's been in place since the first running of the Boston Marathon in 1897 in which students line a quarter mile of the course motivating runners with hoots, hollers, high fives, and, yes, kisses. The cheering is so loud you can hear it from a mile away.

18

THE SILLY AND
THE SUBLIME

Life happens, what matters is how you respond.

Eight and a half hostile miles separated me from the next ves-
tige of sanctity, the Duncan Canyon aid station. While a pit-
tance in ultramarathoning terms, it takes the average American
nearly four days to cover such a distance on foot, and this is
done primarily on manicured walkways that are designed for
easy passage, not unpaved trails through the mountains. This is
a sport of distorted proportionalities, a place where otherwise
physically formidable acts are considered ho-hum. Nothing
about ultramarathoning is in moderation.

Those 8 1/2 miles proved to be most psychologically distressing. I remember running with a guy, Frank, from Van-
couver, Washington, not British Columbia, and trying to keep

the two sorted out seemed to overwhelm my mental process-
ing power, my eyes glossing over into two colorfully spinning
throbbers, the kind you see when a computer is thinking. Frank
was reformed—from alcohol, I think, though it could have been
drugs—and he had traded one addiction for another, albeit a
healthier one. He was a strong runner, I remember thinking,
so he must have been heavily addicted, somehow correlating
the two when obviously no such relationship exists. Or maybe
it does. *Canadians are hardy*, I thought, *but wait, he's not Canadian.*
More spinning throbbers.

When things start to unravel during an ultramarathon gen-
erally two outcomes ensue: either one magically revives, or one
spirals downward to ultimate ruin. Unfortunately, I was trending toward the latter, falling apart one dreadful piece at a time.
Why this race, of all races, was going so badly weighed heavily
on me, and why so early on, barely two rounds into a ten-round
bout, made the burden feel even more crushing. Just keep moving forward, I told myself. Don't think about anything but your
next step. The indigenous Sierra Miwok peoples are said to have
had no word for tomorrow or yesterday. Be like them, I told myself. Don't think about the future, and don't reflect on the past.
Just move in the present moment, the here and now, and take
the next step to the best of your ability. And the next, and the
next . . . Slowly, step by step, I slipped into a trancelike state.

Zombies come in many forms. You have the genetically
mutated zombie, the alien-invasion zombie, the supernatural
zombie, the sharp-blow-to-the-head zombie, the occult-voodoo
zombie, and of course the severed-limb zombie. I was some hid-
eous amalgamation of all of these when I came thrashing into
the Duncan Canyon aid station. Gasps were heard, and terrified
whimpers. People impulsively recoiled, mothers protected their

children, priests thrust crosses stiff-armed to repel the evil. I needed help but no one would come near me. I was a pariah, an untouchable, a vile creature too plunged into the underworld for salvation.

Then someone yelled commandingly, "Get this man water!" A merciful aid station worker had taken pity upon me. Next thing I know she's holding a cup of liquid to my lips. I slurped it, slowly at first, then gulped the balance instantaneously. She instructed the person who'd brought the cup, "Let's get him some more, and also a bandanna with ice."

I slowly tilted forward like a wilting asparagus, arresting my droop only by propping my hands on my thighs, hunched over, eyes to dirt. She unexpectedly began rubbing my back, the way a mother caresses a discomforted child. *Who is this compassionate individual?* I wondered. I was momentarily convinced that Florence Nightingale was working the Duncan Canyon aid station. I glanced sideways.

"Ann Trason?"

I rubbed my eyes in disbelief. Then I looked again. It was! The great Ann Trason.

She looked at me. "Dean Karnazes?" There was a suspended moment of silence. *"What happened to you?"*

I shook my head. "Twenty-four years of living?"

Ann quipped in acknowledgment, "I hear ya there, it all seemed so much easier back then."

It was surreal to imagine the legendary Ann Trason dutifully nursing me back to health at some random aid station along the Western States course—a race she so thoroughly dominated for more than a decade—yet that is precisely what was happening.

"We need to cool you down," Ann conferred, "and get you rehydrated."

I knew these things as well, but not until Ann said it aloud did I take seriously the fact that there was no possibility of toughing it through overheating and dehydration the way I once was able. Despite being in top shape for my age, I was still my age. This reality was confronting me with an incontrovertible immediacy. Trying to deter Father Time is like trying to outrun your shadow, and the lower the sun sets, the longer the shadow that is cast. Escape is a fool's errand. Perhaps I am a fool.

Her helper brought another cup of water and a bandanna filled with ice. Ann tied the bandanna around my neck and started drizzling water on it to activate the cooling. They filled my bottles and handed them back to me. "Let's walk," she said.

I started walking out of the aid station with Ann Trason, a moment I will never, *ever* forget. I was a starstruck fan boy and Ann was humble as apple pie, rubbing my back and offering sage words of guidance. "Take care of yourself," she said. "Stay hydrated and stay cool. You've been here before, you can do this." She seemed legitimately concerned (and I can't say I blame her). "You've been here before," she repeated, patting me on the back one last time. "You can do this."

The high of interacting with Ann carried me about halfway to the next aid station. But then that familiar mental lethargy started creeping in again, cobwebs forming in the hollow recesses between my ears, my eyes incapable of fully opening, light only partially getting through. I'd entered a strange purgatory, not asleep but not wholly awake, time passing in groggy snapshots that were storyboarded together with a weird frame delay. A twist in the trail occurred, and only afterward would it occur to me that a twist in the trail had occurred. I could hear music and commotion coming from the Robinson Flat aid station, though I couldn't tell if I was getting closer or farther away. Sounds were echoing as if bouncing off canyon walls, but we were running through an open forest. Time moved slowly, then quickly, and then slowly again, my eyelids drooping downward, blanketing my field of vision. I needed sleep.

And sleep was the first intention I set out to accomplish once finally arriving at Robinson Flat. It was a noisy and raucous place, being the largest aid station along the Western States route thus far. But I sought solitude, not activity. Behind a row of vehicles off to the side of the course I drifted, finding a suitable place on the ground to crumple over, tucked in the shadow of a large vehicle. I lay in the fetal position in the dirt, left shoulder to the ground, the posture sleep experts advise helps increase blood flow and brain health. Lord knows I needed those things now. I nodded off.

When I opened my eyes it was evident time had passed, though how much time was indeterminate, like coming to after surgery. Oddly, staring at me was a young child. The youngster's presence must have awoken me.

"You okay, mister?"

I blinked several times to clear my vision. It was difficult to decipher whether they were a young boy or a young girl. They had golden locks and rosy cheeks, and they were barefoot. I imagine their name being Cosmos or Willow. My head was on the ground and I looked at them sideways.

"Yes, I'm okay," I said. "Just had a brief nap, that's all."

They kept staring at me with a wondrous curiosity, as though they'd stumbled upon a loose zoo animal.

"Say," I asked, "are those Sour Patch Gummy Worms?"

They were holding something in their hand and they slowly tilted their glance downward to look at the bag.

"Green apple is my favorite," I added.

They held the bag out to me.

"You sure?"

They held the bag higher, as if affirming their willingness to share. So I took it. I gently peeled open the top and started fishing around for the green ones. I put a couple in my mouth and chewed slowing, then zipped the package tight and handed it back over.

They slowly reached for it and when they took hold I jokingly pulled back and wouldn't let go. But they tugged equally hard and finally after some back and forth I laughingly released the bag. They stared at me queerly, then turned and wandered away.

Gradually I got to my feet and walked into the frenzied melee of the aid station. Everyone seemed rushed, yet I felt as though I'd just emerged from hibernation, woozy and out of sync with the hurried pace. I spotted a longtime coach of mine and personal friend, Jason Koop. "Karno!" he bellowed when he saw me.

"I just saw a cherub."

He looked at me oddly. "A what?"

"An angel. I just had an encounter with an angel."

"What on earth are you talking about? Are you all right?"

Scratching the side of my head, "To quote from *King Lear*, 'I fear I am not in my perfect mind.'"

"I'll say. You need to pull it together, man."

"Koop, I think we're beyond that point now."

"You might be right."

"Geez, thanks, coach."

He put both hands on my shoulders and looked me in the eyes. "Karno, you were dealt a lousy hand today. What matters now is how you respond." He lifted both hands and patted me on the shoulders once again. And then spun me toward the exit.

And that is what I love about Jason Koop, his ability to cut through the bullshit and tell it like it is.

Moving to the table before the exit to fill my bottles, one of the volunteers had a copy of my first book. He asked if I would sign it, which I did—weird as signing a book midrace is—and then asked for a picture of us together. He had to be at least six-three, so I stood on my tiptoes to fit in the frame.

"Thanks, Karno," he said with a smile. "How's it going?"

I looked at him. "That doesn't matter. What matters now is how I respond." I think he got the point. "Let me fill your bottles."

He brought them back to me and off I set. An ultramarathon is not a single experience, but a series of little moments strung together in a narrative thread that becomes the complete story in due course. Some moments are silly, others sublime—not that much different from life, really. In so many ways an ultramarathon is a microcosm of life, one compressed into 100 miles of running, the full breadth of humanness is experienced, the physical, the emotional, and the spiritual, and along the path a story gets told.

What I've come to realize is that the ending is not what matters most; sure I may get another buckle today (or may not), but reaching the finish is not the ultimate prize, it's the story that's lived along the way. And today was turning out to be one hell of a story.

19

THE MELTDOWN

Stuck between a walk and a hard place.

If you've ever spent time along sections of the Western States trail you may wonder how this gnarled and rutted single-track trail cut through the Sierra could possibly be traversed by anyone other than the fittest of the fit. Yet the path between Last Chance and Devil's Thumb—some of the most treacherous terrain of the entire course—was once a tollway.

Of course, I'm not talking an interstate; this was during an era well before motorized vehicles. The toll was collected from prospectors on foot and on horseback, by a local couple that owned a nearby hotel and saloon called the Half-Way House. They got tired of collecting dead bodies at the base of the canyon and used the money to improve the path. It helped, a little. But given the severity of the terrain, there was only so much that could be done. People and animals still periodically lost their footing and tumbled to their deaths. Hence the name, I guess, of Deadwood Canyon.

I was standing on this precipice preparing to head down, hoping not to end up like one of the earlier travelers. After 43 leg-powered miles forging mountain and valley, the problem with running downhill is that, frankly, it really fucking hurts. Like, kills. To me, it's not the 18,090 feet of climbing at Western States that wrecks you; it's the 22,970 feet of descending. Running downhill is broadly classified as an eccentric muscle activity, and every time your foot contacts the ground microscopic filaments of muscle fiber get a little jujitsu kick. Over time your quads essentially become like a slab of meat on Paul Bunyan's chopping block, only instead of an ax he's wielding a wooden mallet.

There's a term that gets tossed around in gym circles: DOMS (no, it's not a fancy champagne). DOMS stands for Delayed Onset Muscle Soreness. It's the feeling you get the morning after you decided it was a good idea to lift some weights with a team of rugby players, or play a few sets of tennis with Roger Federer. Getting from the bed to the toilet the next morning has never been a more excruciating test of fortitude.

When it comes to ultramarathoning, one might argue that the delay between the damage and the onset of soreness takes place during the event itself (a significant enough duration certainly exists between the two). Maybe that's splitting hairs, but all I know is running downhill from the Last Chance checkpoint to the pit of Deadwood Canyon was like having dental surgery without Novocain. A pain so intense it made me puke. Out it came.

I wiped my chin with the back of my hand and remembered that my wife's a dentist and so I probably shouldn't be making analogies between pain and dentistry. Bad karma. Especially if

I needed to have some work done. Which, unfortunately, was becoming more frequent these days.

Seems those sugary energy boosting gels are not entirely simpatico with teeth. In fact, if a dentist wanted to devise the perfect recurring revenue scheme he would concoct a tearable pouch (usually done with one's teeth), fill it with sticky goo, and advise people to consume it at regular intervals throughout the day. Even better, throughout the night as well. And there you have an ultramarathon (and one very employed dentist).

Though I needn't concern myself with that now. Hungry as I was, the thought of consuming another gel repulsed me. All I'd been eating the entire day were sweet and sugary finger foods, and now I was suffering the repercussions: nausea and light-headedness. At the floor of Deadwood Canyon is a narrow, swinging footbridge. I felt like a drunk trying to cross it, flailing side to side, trying to stabilize myself using the slender cable guardrails. One step forward, a wobbly half stride sideward, and then another step forward.

On the far end of the bridge was a small spring trickling down from the canyon wall. I leaned against it backward and tilted my face toward the sky to let the water run over my forehead and chin. Although the flow was little more than a drizzle, it felt exceptionally pleasant. So much so I thought I may spend some time here.

Unfortunately, my blissful solitude was interrupted by another racer, a rather vociferous one.

"Feelin' it, are we?" he guffawed.

"Ooh, you could say that."

"Well, get ready for the climb. It's 2.08 miles, there are eighteen hundred and twelve feet of ascent, and a total of thirty-six switchbacks," he said chirpily.

I looked at him open-mouthed and belligerent. "When I signed up for this race I was told there'd be no math."

He gave me a strange, befuddled stare.

"Yah, well, I'm going to keep my heart rate under a hundred and sixty. It's a doozy. Good luck, buddy."

Off he went.

Eventually I gathered enough energy to confront those thirty-six switchbacks. He was right: not one more (thankfully) or one less (regrettably). The climb deposited me at the Devil's Thumb aid station. It couldn't have come sooner.

Fortunately I landed in capable hands. The Devil's Thumb aid station has a long-standing and committed crew of volunteers and they are some of the nicest folks you'll ever meet.

Two women smilingly greeted me as I came limping in, visibly damaged and sore.

"This race is a motherfucker," one of them says to me.

"For crissakes, Louise," the one standing next to her rebuked, "where'd you get your manners, on a pirate ship?"

She then turned to me. "My apologies. We don't let her out of the barn very often."

Such sweethearts, I thought.

"What can I get for you?" the polite one asked me.

"A new set of quadriceps would be nice."

"We can do that." She put two fingers in her mouth and dispatched an earsplitting whistle. "YO! Sheila, get those power tools over here."

What arrived can best be described as a cross between a handheld jackhammer and a masochistic sex toy, and it made a high-pitched shrill like an industrial-grade bug zapper.

"Fifty thrusts per second." She held it in front of her to inspect the conical head.

Pirate mouth peered over the polite one's shoulder to marvel at the machinery. "Why can't I find a man like that?"

"Would ya get outta here!" The polite one shushed her off with the back of her hand. "Go on, scram. Go find Captain Hook."

She turned back to me and took aim at my lower torso with the twenty-four-volt battering ram held stiffly in front of her.

"Wait, wait . . . I'm not sure." I held my hands up in reluctance.

She nodded. "Okay then, let's start with the glutes. Turn around."

"Um, not to seem ungrateful, but there is no chance in hell I'm letting you come at me from behind with that thing."

"All right, suit yourself." She took aim at my front side.

"But . . . but . . ." I squirmed in trepidation.

"C'mon, you know you want it," she said, narrowing in with a devilish glint in her eye.

"Go slow . . . pleeeease . . ."

It was too late.

"AAHHH!" I yelped in agony as ten thousand karate chops pummeled my tender fascia.

"And the hip flexors." She moved to the top of my thigh and it felt like a jolt from a cattle prod. I flinched and convulsed as another wave of nausea welled up in my stomach.

She looked at me genuinely. "You really are sore."

"You don't even know," I said slowly, massaging my temples.

She lowered her vibrating revolver. "You've been down this road before."

"You mean literally or figuratively?"

"Both."

"Yes, I have. And I'd sure like to continue going down this road, but all things come to an end. Maybe it's time to join AARP and take up bridge."

She put down the robogun. "You seem more of a gin rummy man to me."

I snickered. Then she struck a more sincere tone. "You've influenced a lot of us here today, more than you will ever know."

Her remark surprised me, as such comments always do. To me, it was the other way around. "And a lot of you have influenced me," I replied. "More than you will ever know."

With that I turned and made my exit. I didn't fill my bottles, didn't try to eat, and didn't bother sponging down. Rookie maneuvers, evidence I was getting increasingly reckless. Absolute exhaustion does that to a man. At least my quads were a bit looser.

The plunge from Devil's Thumb down into this next canyon was protracted and unbearably oppressive. It is one of the longest descents of the course, and the heat and humidity were working their unique combination of wickedness the deeper we dropped. Near the pit of the gorge the trail passed Deadwood Cemetery, the resting point for many men who came seeking their fortunes only to meet an untimely and strikingly permanent end to their dreams. *Whose dreams will meet a similar fate today?* I wondered. *How many racers won't reach the finish line?*

At the bottom of the canyon there was a creek. I crawled in. Completely. The water enveloped me, entering my shoes and socks, washing over my hat and head. I lay in the stream not wanting to get out, not wanting to face the hellish climb up the far side of the canyon to Michigan Bluff. It didn't seem possible. I wanted to float downstream, float away and be forgotten. And that is when I knew I would not finish the race. My fate was sealed: my dream had become my nightmare.

At some point after a long and mournful soaking I crawled out of the river. I'd stopped checking my watch, so who knew how long it'd been? When I resumed progress on the trail the dirt and debris stuck to my sopping shoes and I left a track of water like a wet dog. There were rocks and pebbles inside my shoes and socks and I knew this was damaging to my feet. Knew, but did not care. I was defeated. My leadfooted steps up the ascent were little nails in my coffin. There was no running left; I was stuck between a walk and a hard place. Midway up the climb my hometown mate and running pal Karl Hoagland came upon me. His pace was steady and calculated, meticulous and well executed. We chatted. He was on course for his tenth finish, and his spirits were high. He urged me onward until he realized I was past the point of no return, whereupon he respectfully left it at that and resumed pursuit of his own goal. I watched his silhouette fade into the distance until there was nothing more. And I was left alone in the dust, feeling my age.

20

HEAD FAKE

To rise from the ashes first you must burn.

Eventually the climb was contended with. It was unclear precisely how long it had taken, but it was certainly a pitiful duration. In the final stretch of course leading to the Michigan Bluff checkpoint the trail widened and turned slightly downward into a perfectly groomed pathway ideally manicured for a brisk resumption of running. I walked, head slung low, defeated.

Music and voices coming from the aid station could be distinctly heard before rounding the last corner, probably because I was walking so slowly there was little else to do other than focus on the clamor. When I finally did arrive at the checkpoint I was weighed—a mandatory requirement for all runners—and then allowed to pass through the medical area to access my crew. The first person visible was Nicholas, his broad shoulders and solid build quite distinguishable.

"Hey, Pops." He looked at me. "How's it going?"

His simple question catapulted me into a contemplative look forward, a fortuitous vision forming of my future as I pondered the consequences of my next action, the upcoming scene playing out in eerie dreamlike animation, as though fate was giving me a signal of the potential outcome of my next decision. I looked at Nicholas and opened my mouth but no words came out. Instead, I watched my life flash before me like a grainy motion picture.

"It's not going so well," I informed him.

He inspected me more thoroughly. "Follow me," he said, "we've got everything set up for you." He turned.

"I'm not going."

He turned back. "What?"

"I'm done. I'm dropping out."

"What? I can't believe I'm hearing this. You always tell people to never give up."

"It's not my day."

He was growing angered. "You don't qualify it with, 'It's okay if it's not your day.' You tell people to *never* give up."

"Maybe I'm a hypocrite."

"Damn right you're a hypocrite!"

People turned to look.

"Shhh . . . you're causing a scene," I said disapprovingly.

"Don't tell me what to do!" And he stormed off.

I turned and the many faces watching the squabble abruptly looked away. There was an uncomfortable silence. A volunteer approached me.

"I think your stuff's set up over there." He pointed up the road.

"Thank you," I said. "Listen, I'm really sorry about that."

He didn't respond, just turned and walked back to his post.

With my head lowered I set out up the road. Eventually I saw my mom.

"Hi, honey! Come on over, we have everything set up for you. Where's Nicholas?"

Just then Nicholas appeared. "Let's go," he said hostilely. My mom looked at him, concerned.

"Nicholas," I said, "can you please give me a few minutes? I've been running for 55 miles."

"I don't care. I've been driving for five hours. I want to be with my friends."

"What's the matter?" my poor mom asked.

I looked at her, trying to be reassuring. "It's not my—"

Nicholas cut in. "He's quitting."

Now she looked upset. "Is everything all right?"

"Let's go!" Nicholas said to me angrily, too ashamed to look at his grandmother. He stomped away toward the car.

My dad came rushing over. "What's wrong?"

"We're leaving."

He looked at me in utter dismay. Then the car horn started blasting.

"That's Nicholas. He wants to go."

"Go?" my dad asked. The horn blasted again.

"Here, let us help you." My mom started hastily gathering everything they had carefully laid out for me. But the horn kept blasting.

"Can you just collect it all?" I asked.

"Sure, honey, that's fine." She looked so sad. Dad stood beside her, and his face was ashen and withered. The horn blasted more. I turned and walked away.

Getting in the car was brutally painful. My muscles were aching and stiff, so I tried lowering myself using the door handle.

"Come on, get in!" Nicholas demanded. "What's taking you so long?"

When I finally lowered myself onto the seat he irately jammed his foot on the gas pedal and the door slammed violently shut, nearly crushing my fingers in the process.

"Nicholas, will you please take it easy?"

"You're pitiful," he huffed.

The road was twisting and snaking and he drove like a maniac. I wanted to unlace my shoes, but I was being tossed about too aggressively to attempt any movements. There was an acerbic taste in my mouth and I felt nauseous.

"Could you please slow down?" I asked politely.

"Don't tell me what to do!" He drove faster.

Something rolled out from under the seat and hit me on the ankle. I reached down and dug around. It was a bottle of Jack Daniel's.

"Nicholas, have you been drinking?" I said sternly.

He didn't respond.

"I asked you a question. Have you been drinking?"

"What's it to you?" he scoffed jeeringly.

"You're my son."

"Since when have you been so interested in *me*?"

"You're my son."

"Oh, bullshit. Everything's about you and your running. You don't care about anyone else."

"That's not true. I love you."

"Like hell you do," he said with a sneer. "It's all about you!"

He turned on the radio full blast and didn't say another word to me the entire drive home. When we arrived he pulled up in front of the house.

"Get out."

I wearily got out of the car, feeling arthritic and aged. I turned to look at him, but he stepped on the gas and sped off down the road in a cloud of dust.

Dragging myself up the driveway, my feet were reddened and swollen, stuck in my shoes. I felt sunburned and blistered. Julie heard me coming through the front doorway.

"What are you doing here?" she said in complete shock.

"Things didn't go so well."

"What does that mean?" she questioned.

"That means I dropped out."

"What? How did you get home?"

I breathed heavily. "Nicholas drove me."

"Where's Nicholas?" She was starting to get upset.

"I don't know. He sped off."

"Sped off? What do you mean, *sped off*?"

"I'm not sure." I shook my head. "We got in an argument."

"And you let him drive off like that? You didn't try to resolve it?" Now she was dismayed.

"I tried." I looked at the ground and then back up at her. "I tried."

"So you dropped out of the race, got in a fight with your son, and now you're home?"

I nodded my head sadly in agreement.

"I told you to return with your shield, or on it. You've come back with your shield between your legs."

"I tried."

"Tried?" She looked at me in disgust. "Pathetic." She turned and walked away.

My world had collapsed with an abrupt finality. The ending to my running career, to my love and vocation, was a disgrace, as every quitter's life is. I sat on my front stairs, wrapped my arms around myself, and started rocking back and forth. I squeezed my eyes tightly and rocked back and forth.

"Dad," I heard a voice, "Dad! Is everything all right?"

I felt a hand on my shoulder rocking me back and forth. "Can you hear me? Dad, you there?"

I blinked several times to clear my vision, the nightmarish daydream I'd just had dissolving into thin air. It had all been an illusion, a prophecy, imagined, not real.

"I asked you how's it going?" Nicholas repeated his question from earlier.

When I opened my mouth this time the words did come out.

"Sorry, Nicholas, I was experiencing a strange moment of darkness. But it wasn't real, thankfully. Just a warning sign. I've risen from the ashes."

"Huh? You okay?"

"I am now," I let him know. "And I'm definitely not stopping, that's for sure."

"Good to hear. Follow me, I've got everything set up for you."

Just then a volunteer approached.

"Hey, Karno, is this your son?"

"Oh, howdy, Peter. Yes, this is Nicholas."

"Big kid." He held out his hand.

"Nicholas, this is an old friend."

They shook hands. "Helluva handshake, too."

I chuckled. "Unlike his svelte father, he's a football player."

"Nice meeting you, Nicholas."

"And you, Peter," Nicholas responded.

Peter turned to me. "Good luck, Karno. Looks like you're in able hands."

I thanked him and then Nicholas led me down the way.

He was right: everything was set up perfectly for me. There was a collapsible chair open and waiting, and in front of it he'd laid out a big towel, my gear bags unzipped for easy access. My mom and dad were standing by, waiting.

"Hi, honey! Come on over." My mom gave me a big hug. "We've been having such a nice time catching up with Nicholas."

"Can we get you anything to eat?" my dad asked. "We've got some of that raw almond butter and a spoon."

"*Yum*," I sputtered.

"I'll go get it," my mom said happily.

"Mom." I stopped her. "I'm being facetious. All I've been eating today are spoonable foods. My stomach turned south a few miles back. I'm craving something more substantial."

I sat down in the chair.

"How does Mexican sound?" Nicholas said.

My head snapped. "*Mexican?*" I said to him slowly, "Nicholas, I would eat a raw iguana right now."

"I've got an extra burrito in the car," he said casually. "I'll get it."

"You're serious, aren't you?"

"Yeah, Dad."

"Where on earth did you get Mexican food?"

He looked at me strangely. "There's a Taco Bell in Auburn. You've been spending too much time out in the wilderness."

He came back with the burrito.

"What happened to your Paleo diet?" my mom asked in shock.

"Fifty-five miles of Western States happened to my Paleo diet, that's what." I motioned to Nicholas. "Gimme that thing."

When he handed it to me I hyperextended my jaw and stuffed the entire mass in lengthwise, getting about three quarters of it down the gullet. Then it got stuck.

"Giv me somp tin ta drik," I gurgled out the corner of my mouth.

"Dad!" Nicholas said, appalled. "What happened to your manners?"

"Iffie ive mils r wetern staes appen ta mye maunrs," I muttered.

Another runner ran past and waved. I looked like a python with the hindquarters of a partially digested guinea pig sticking out of its mouth. I waved back.

"Would you like me to change your shoes and socks?" Nicholas asked.

I nodded in the affirmative, and then tilted my head backward to let gravity help drive the burrito down the hatch, jerking my head violently several times to assist passage.

Nicholas unlaced my shoes and removed them, exposing my porcelain white feet that were shriveled as prunes. He gently toweled them off and replaced my socks with a fresh pair. Slowly he slipped one foot into a new pair of shoes, then the other, and then expertly laced them, overall doing a much finer job than I could have done on my own. Quicker, too.

"Feeling better, Pops?" he asked.

"Infinitely so, thanks to you."

I slowly stood up and prepared to depart.

Nicholas had two fresh bottles waiting.

"What's in . . ." But he interrupted me.

"The right one's got Perpetuem, the left one Heed. And there's a packet of Endurolytes in each stash pocket."

He'd filled them with my preferred products.

"Thanks, man, you're so on it." I turned and started to leave.

"And don't forget this." Nicholas handed me the running pack I'd asked him to bring. Had he not reminded me, I would have forgotten it. I was seeing a different side of Nicholas. He had a casual efficiency to his actions, and he didn't let details go unnoticed. He was bright and perceptive. Meanwhile, his father would be hard pressed to write his name in the dirt with a stick.

"I'll see you guys in Foresthill," I said to them, "and thank you for being here."

21

EMBRACE THE SUCK

To know thyself one must push thyself.

I think we run 100 miles through the wilderness because we are changed by the experience. What takes a monk a month of meditation we can achieve in twenty-four hours of running. With each footstep comes a slow diminishment of self, the prickly edges of ego whittled down until something approaching the divine emerges. Even during a race with no shortage of human folly, great moments of clarity are achieved. Running an ultramarathon builds character, but it also exposes it. We learn about ourselves, we gain deeper insights into the nature of our character, and we are transformed by these things. To know thyself one must push thyself.

And we do. After leaving the family in Michigan Bluff the course once again descended, this time into the bowels of Volcano Canyon—yet another prehistoric gouge in the earth's crust—where the moist, humid air was stagnant and syrupy, the sun beating down with searing intensity. The trail now wove along a narrow rocky pathway that was overgrown with branchy vegetation, some of which entirely obstructed the ground, making the footing rather tricky at points. My pace was slow and deliberate, mindful of the potential hazards, but still I failed to notice a root obtruding from the earth and kicked it squarely with the big toe of my right foot.

The outcome was immediate and violent; my upper torso came crashing to the ground with such force it kicked up a massive spiraling cloud of chalky dust like a sack of flour toppling off a delivery truck and rupturing on the asphalt. I lay dimwitted in the dirt, my senses knocked from me, and a droll sound repeating in my ears, *"Cuckoo, cuckoo . . ."*

My head faced forward, my chin nuzzling out a little divot in the soil. I watched as a bug walked in front of me. It had two bulbously protruding eyeballs and a rust-colored thorax with translucent wings tucked to either side. It scurried about anxiously, as if there were much work to be attended to. Another bug of the same variety appeared on the opposite side of the trail. It busily scampered over to its kin, the pair of them dancing energetically in circles with each other, and then both promptly scuttling off down the path merrily together. *Perhaps many baby bugs would come of the union*, I thought.

I remained there for quite some duration waiting for someone to find me and help get me to my feet, but that wasn't happening. It amazed me how spread out the field had become, how infrequently I encountered other runners. Then again, this was a 100-mile footpath. It should come as no surprise how thin the pack had become. With no one emerging for assistance, I pulled myself upward and rose to my feet on my own. My big toe throbbed and I wondered whether it had been broken or dislocated. With certainty, I knew the toenail would soon be departing. *Embrace the suck,* I told myself, *toenails are so overrated.* The first few steps were excruciating, though in time the pain subsided. I think the neurons get so flooded they eventually give up. *Ah, fuck it,* they say. Give the guy a break.

Foresthill is the most accessible checkpoint of the entire race and subsequently the busiest. It is one of the few brief sections of the course where the trail exits the wilderness and proceeds along a paved road for a short stretch. Many more people lined the roadways here than anywhere else along the route. It wasn't the Chicago Marathon, but having any cheering fans along the course during an ultra is a welcome sight.

As I made my arrival it was hard not to feel self-conscious. The dirt I'd frolicked in during my recent tumble had become one with my clothing, the damp fabric absorbing the powdery brown soil into its textile matrix. Had a race number not been pinned to my jersey, or had this been any other backcountry rural community, it would've been easy to mistake me for a wandering vagabond. But this was Foresthill on race day, and people clapped and cheered despite my soiled wardrobe and haggard appearance. To an outsider this must look like an exceptionally entertaining freak show.

My crew was waiting for me along the roadside. Nicholas had once again spread everything out in uniform fashion for easy access. The kid was amazing me.

"Thanks, Nicholas," I offered, "this is perfect." I looked at him more closely. "Your cheeks are so red."

"It's hot," he said, panting. "I don't remember it being this hot before."

"Do you remember any of it? You were just a kid."

"Oh, yeah, I remember having a squirt bottle and spraying you off."

"Well, you also spent a good deal of time spraying your sister off, too. And she shot back! That's probably why you don't remember being so hot. That and global warming."

Just then someone boomed, "KARNO!"

It was my pacer and longtime companion in mischief and adventure, Topher Gaylord. He and his wife, Kim, had been chatting with my parents. They were old friends.

"Let's go!" he commanded.

"Give me a sec, Toph. I need to regroup."

"Whaddya mean? You've done this race a dozen times. Let's go."

"Easy, cowboy. You've got a trophy room filled with buckles yourself, and you know that don't mean diddly-squat today. This race don't care about past glories, you know that. In fact, didn't you DNF at this spot last time?"

"No. It was at Michigan Bluff."

"That's even worse! Michigan Bluff was 6 miles ago."

"Just sayin' . . ."

I reached over to flick his forehead, but he recoiled the way a boxer does eluding an opponent's punch.

"Hey now," he reprimanded me, "I'm here to help you."

I tried again, but he dodged my flick once more.

"All right, you two, break it up!" Kim Gaylord stepped in.

"He started it!" Topher pleaded like a child. Mind you, the man's a prominent executive in the outdoor industry and member of the Western States board.

Nicholas handed me some fresh bottles and asked if I needed anything else. "I could use a cattle prod in case Topher gets unruly."

"That's funny," Nicholas said.

"Who's jokin'?" I grinned.

Kim handed us our lights. "Don't forget these."

This was the first time ever I needed to take lights at the Foresthill checkpoint—mile 62—so early in the race. Usually it was many miles down the trail before darkness fell. Not today.

"Do you have extra batteries?" Dad asked.

We looked at him incredulously. "Pops, remember, they're rechargeable these days."

Out of sync with the times as the man was, his comment carried an endearing undertone of nostalgia. Most of the racers and crew here today probably had no idea that headlamps once required batteries. Back in the 1990s, Topher and I attempted a speed record on the John Muir Trail—now colloquially known in ultra-speak as an FKT* on the JMT—and the excess weight of spare batteries was a real concern. Burdensome as carrying spare batteries had been back then, those bygone memories of yesteryear never failed to invigorate the spirit.

With fond reminiscences and fresh bottles, it was time to mosey along. "C'mon, Toph. If I stay here any longer I'll rust."

"Attababy, Karno!" he responded with characteristic zeal. "Let's get after it, for ol' time's sake."

The plan was for Topher to pace me to the Rucky Chucky River Crossing and Kim to pace me from that point to the finish. Of course, that was just a plan. There are no guarantees in this sport.

We said farewell to the others and headed off into the wilderness, the setting sun turning the early evening sky into a kaleidoscopic fireworks display. Deeper into the wilds we proceeded, the colors and textures intensifying more and more as we ran. I'd just spent from sunup till sundown in the outdoors,

---

* Fastest Known Time.

something fundamental to our origins yet nearly unheard of today in our developed world. The experience of being outdoors all day in nature is life-affirming in some vitally human way, and the ultramarathon gives us this forum for inner enrichment. The air was motionless, few sounds could be heard other than the distant American River snaking down the cavernous valley. The tranquil swish of the river and the richness of the tones filling the sky were spellbinding, mesmerizing us in awed hypnosis. Moments like these cannot be adequately described; they must be lived.

"I love this, Topher," I said. "I don't ever want this to go away."

"Me, either, Karno."

You can get enough of most things in life, but never nature.

We continued descending into the American River Canyon paralleling the tributary, the reverberation of rushing water at times distinct and clear and at others subdued and gentle. Known as California Street, the trail in this section consists of a series of undulating rises and descents, none particularly insuperable but in their totality rather wearing. The flats and descents could be run, but the ascents, brief as they were, reduced most to walking.

Topher was a good sport, and that could be said of most pacers. An elite runner himself, his role today was essentially providing moral and emotional support.

"Gadzooks, Karno, what's wrong with you? We're gonna be out here all night at this pace. Pick it up, bro."

He was beyond striking distance, and in my dilapidated state there was no way I could possibly catch him. I swiped at him a couple of times, but it was no use.

"You're really suffering, aren't you?"

"If I could somehow catch you you'd be suffering, too."

"Be nice, fella. Here, I'll share with you."

"Whaddya got?"

He held out a pouch of chocolate-covered espresso beans. It was an exceptionally equitable peace offering and I gladly accepted his gesture of goodwill. Munching on a couple of them I muttered, "Thanks, Toph. These are good." Rare are the friendships that evolve into a deeper brotherhood where the full range of human emotions and fallibilities could be so openly expressed. Running great distances together stripped away the protective coating and allowed a more honest melding of souls. Those friendships forged in footsteps were the best.

We shuffled along steadily, the luminous alpenglow leisurely evaporating into the distant horizon, a blanket of darkness moving over earth like a slow eclipse. In due time it all faded to black and we became nocturnal creatures. With headlamps switched on and illuminating the trail, reality was reduced to the reaches of a shaft of light. At points—vistas unseen in the darkness—a beaming lodestar could be seen in the distance, the lights of the Rucky Chucky River Crossing checkpoint. It seemed close at moments and then impossibly distant the next. I'd never traversed this section of the course in darkness before, and the experience was disorienting and jarring to my senses. So close the checkpoint seemed, yet so far away.

Just then the lights of other headlamps could be seen and the sounds of footsteps coming up behind us could be heard. Like the beacon ahead, it was difficult to ascertain if the party was directly upon us or some distance off in our lumbering slipstream.

Eventually the voices became more distinct and it was clear their proximity was near.

"Ello," someone said in a thick French accent. I recognized the voice; it was that of Fabrice, the runner I'd shared some footsteps with earlier in the day when he had been struggling.

"You're back," I said in acknowledgment of his revival.

"*Oui, monsieur,* I have doubtful memories. Many linger."

It sounded complicated to me, too much to unpack at this trailside hookup. "Look," I said, "soon we'll be at the river and my friend Topher will anoint you with water."

"Oui," Fabrice uttered once again, rubbing his forehead.

When he and his pacer were out of earshot, Topher laid into me: "Whaddya talkin' about, anoint him in the river?"

"Need I remind you you're an ordained minister?"

Allow me to explain. In 2006, prior to running fifty marathons in fifty states in fifty days, I'd asked Topher if he would do a brief renewal of my wedding vows with Julie before the start of the first marathon. I figured if I died or she left me, she'd have to divorce me twice. Not a man of half measures, Topher got very invested in his pastoral duties. He took an internet course through the Universal Life Church and got an online diploma of priesthood. For the occasion, he rented an ornately embroidered clerical gown and conducted himself like a devout man of the book, all principled and sanctimonious. He later had his diploma framed.

Of course, I've never let him live it down. I once asked his holiness to bless my blackened toenail so that when it fell off I could conduct a proper burial. He was a decent chap about it, but still took his clergyman status very seriously, his internet diploma as verification of his loftiness. "All hail!" I joked, though he seemed to view my torment as a test of his devotion, which served to further elevate my ridicule. Yes, we were good friends. Friends not old enough to forget the past, but friends who knew the past would soon forget us. Reminiscences had become more commonplace than talk of things to come. We were older now, life had moved along. But we had some damn good memories.

The soil alongside the river was sandy and wet, and it

squeaked and crunched underfoot as we marched along, the soft, pliable surface absorbing more energy than it returned, making for strenuous running conditions. Complete darkness now shrouded the trail, flashes from our headlamps ricocheting off the tree branches and gangly vegetation lining the passageway, the sharp, wooly scents of coffeeberry and coyote brush hanging indolently in the air. Temperatures remained oppressively warm, even now on the backside of midnight.

We didn't say much as we ran. Crickets occasionally chirped, their stridulations coming from all directions like surround sound. The occasional splash and churn of the flowing American River made it known that something substantial was present nearby, something fluid, dynamic, and powerful. Although it remained unseen, its force could be felt through your skin, its undulations and countercurrents giving off pulses of energy that moved through your body like sonic waves. We ran along the river entranced in the grand harmony of nature, experiencing the sounds and sensations of earth as it was when we first came onto it. In moments like these there is a primordial connection to who we are that somehow transcends the frivolity of everyday living and takes you back to a place that is simple and pure and elemental to being human. We ran together alongside the river in the darkness, and being alive felt as it should.

When we arrived at the Rucky Chucky River Crossing aid station we were strangely at peace. "Thanks, Topher," I said. "That was real."

He grinned in acknowledgment and drifted away. No further words need be spoken.

22

LONDON CALLING

Desperation is not a plan.

Crossing the river under cover of night was something new to me; never before had I arrived at this juncture without sufficient daylight to make safe passage. Lowering myself down the darkened riverside embankment and cautiously wading into the water, it was unnervingly cold and bracing. The American River was mostly fed by snowmelt from the higher elevations, and to the uninitiated, crossing it could be catastrophic. To those unlucky few, the Western States journey ended at this point when their muscles seized up upon exposure to the whirling, cold-water torrent.

Thankfully, some of us found the occasion just the opposite, renewing. I submerged fully in the chilly liquid, then jumped up and shook vigorously like a wet dog. "Brrr!"

It felt so good I did it again. Once sufficiently doused and thoroughly chilled, I began the crossing. A line was strung across the waterway for safety, and I held tight as I stepped farther into the depths, the waterline rising over my waist. I thought about other races and how Western States compared. To a runner at, say, the Boston Marathon the idea of forging a river midrace would seem preposterous, unimaginable. But here I was, 78 miles into a 100-mile footrace grasping a flimsy rope for dear life trying to avoid being swept downstream. If marathoning is boxing, ultramarathoning is a bare-knuckles bar brawl.

When I reached the far side of the river, I had a pedicure (in my dreams). In reality, I marched up the sandy embankment in my soaking wet shoes and socks and continued onward. Many veterans say Western States begins once you've crossed the river. I'd now done so. The race was on.

Unexpectedly, waiting on the shoreline stood Nicholas.

"What are you doing here?" I was startled to see him; this was not something we'd planned.

"Thought I'd check on you, to see if everything's okay."

"That's nice, but how did you get here?"

"I walked down from Green Gate."

"You realize you're gonna have to run back."

"They told me most people hike this section."

"Most people do, though I might crawl."

"Still not feeling it?"

"Nicholas, there are good races and there are bad races. Let's just say this is not a good race."

He smirked. "Dad, you're running 100 miles. To most people that says enough."

His perspective was healthy. Sometimes when you're so close to something your viewpoint distorts. Fast or slow, everyone out here was running 100 miles. Enough said.

We started hiking the roughly 2 miles up to Green Gate and

what had been told to Nicholas was indeed truthful, most people do hike this section, principally because it's nearly straight uphill the entire way. You could run, but you probably wouldn't get there much faster, and you'd almost certainly expend more energy. Until you can sniff the barbecue pits at the finish line, conservation of energy is always a factor during an ultramarathon.

"How's everything?" I asked Nicholas as we hiked along.

"It's nuts what you guys do."

"Does it seem different now?"

"When I was young, I thought everyone did this stuff. Now I see it for what it is."

"Nuts?"

"I could use other words."

We both laughed. "How are your grandparents holding up?" I asked.

"It's been fun hanging out with them. You know Gramps, he makes friends with everyone."

"I'm sorry, he can be a bit overwhelming."

"I don't mind. You grew up with him so I think it's different for you."

We continued power hiking up the steep incline, Nicholas and I, the stars beaming brilliantly in the never-ending cosmos. And in that brief moment our hearts beat as one, and I felt a paternal connection to Nicholas in a way I hadn't since he was a young boy.

"Thank you for being here," I said to him.

He didn't answer; didn't have to. He felt it, too.

There is a threshold at which words and thoughts cannot cross, a place of deeper awareness and meaning. Such instances are rare, profound, and precious. We continued hiking into the darkness, the two of us, all sense of time and place vanished. Two souls drifting along together in the night.

The first detectable sounds we heard as we drew nearer to

Green Gate were those of my father joking and cajoling with the aid station volunteers. Nicholas and I looked at each other and chuckled. Even at eighty-two, there was just so much of him. It was the middle of the night and the man was only now coming into his own.

The Green Gate aid station is remote and accessible exclusively by foot; thus there were fewer people here than at other aid stations. My dad was talking with one of the volunteers, I think about golf, a lifelong passion of his, and he spoke with great bravado. Yes, it must certainly be about golf.

"Dad," I said, "maybe keep it to a dull roar. We could hear you half a mile down the trail."

"Let him be," the volunteer said in Dad's defense. "He's good entertainment."

"Has he started to dance?" I questioned.

"Wait, it gets better?" the volunteer inquired.

"Don't tempt him," Mom rebuffed coyly.

Just then I saw Kim Gaylord, Topher's better half. Kim was tan and fit, with long, flowing dark hair, and she was like a locomotive. We ran together often, and when she put it in gear there was no stopping her.

During the aid station transition Nicholas had refilled my bottles with Heed and Perpetuem. He was working behind the scenes, and I'd hardly noticed his absence.

"Here, Dad," he said, handing them back to me. "Stay fueled."

There was a pleasant, pub-like feel to the Green Gate aid station, and it would have been easy to relax and stay awhile. Of course, that would be a terminal mistake. The clock was ticking and only so much time remained for me to reach the finish line before the hourglass emptied. I was 80 miles in and had thankfully crossed the river. Now came the hard part.

Kim and I started down the trail together. At this point my pace had speed restrictions placed on it, like one of those go-carts you rent at a county fair. Even with the gas petal flattened to the floorboard, the thing would only go so fast. Calculating my time and tempo, there wasn't much margin for error. Though, admittedly, my mind was incapable of crunching numbers at this point.

Not Kim's. She was pragmatic and efficient, someone very capable of getting shit done. I've seen her operate on many occasions, and no task stood a chance. What's next? Done. Sometimes I wondered how Topher got through the day without her. Then again, sometimes I wondered how I got through the day without Julie (and I had absolutely no idea how my dad got through the day without my mom).

"Okay, what's the plan?" she asked.

"Plan? I dunno, maybe keep heading west and cross my fingers?"

"Karno," she said in a no-nonsense tone, "desperation is not a plan."

We ambled along a bit farther and I could tell her mind was at work.

"Okay," she said, "let's get to the Auburn Lakes Trailhead checkpoint before three thirty."

"Ah, how far is that?"

"Karno, you've done this race before, right?"

"Is it, like, 10K?"

"A little less, but close enough."

"How much time do I have?" I was already assuming the helpless male codependency role. I had a watch on and could easily check for myself, but noooo. Let's face it, fellas, women run this world.

"You've got over an hour."

"Piece of cake. I'll get there way before three thirty."

I arrived at three twenty-nine.

The distance had taken a physical toll on me, but the emotional anguish it inflicted was devastating. What was I doing here? My presence was contemptible. My body was haggard, my steps labored and painful. There's nothing noteworthy of my presence, I felt, I'm simply occupying space on the trail that belonged to a more deserving athlete. I've got my war chest of buckles, so what right did I have continually taking from a sport that had given so much to me? "Brief is the season of a man's delight," the ancient Greek poet Pindar had noted. My glory days were over; if I couldn't somehow contribute to the greater good this was nothing more than a selfish chest pound. Anyone could take, a true champion gives.

Auburn Lakes Trailhead is one of my favorite checkpoints along the route. The place is brought to life with colorful Christmas lights strung along the pathway, and the volunteers dress in Hawaiian shirts and grass skirts and wear leis. It's a festive place, a place of merriment, though my mood was sorrowful.

"I'm going to dump the dirt out of my shoes," I told Kim.

Walking over to find a chair near a darkened back corner of the checkpoint, I came upon a makeshift infirmary. As I navigated around the bodies to take a seat, I observed some of the deadest alive people I'd ever seen. Many were wrapped in blankets and lay curled sideways on the tarp, perhaps asleep, perhaps unconscious. Some sat hollow-jawed with a distant, forlorn look in their sunken eyes, salt crust ringing their faces, lips cracked and dry. It smelled vaguely like a kennel.

I spotted an empty seat and sat down next to another runner. As I did so, I heard him speak. He had a thick British accent.

"I know you."

"Oh. Hey," I returned his greeting, briefly glancing his way. I wasn't particularly interested in conversation. I reached down to unlace my shoes, but he continued.

"My name is Simon, from London."

I sighed. "Hello, Simon," I said while still looking down, untying my laces.

"I met you five years ago at Hyde Park."

"Cool," I mumbled.

"You led a run around the park and I was one of the participants. There were a lot of people."

I kept my head down. "Great. Thanks for joining."

"You signed my book afterward."

"Cool," I offered curtly, still focused on undoing the laces.

"Do you know what you wrote?"

"No, Simon," I huffed, "what did I write?"

"You wrote, 'See you at Western States.'"

I stopped fiddling with my shoelaces. Shivers ran down my spine. I sat up and looked at him for real, a haunting stillness in the air.

"That's . . . that's quite a coincidence," I said, stunned by the fortuitous words I'd written five years ago.

"I just wanted to thank you," he said.

I wasn't sure how to respond; I was at an absolute loss for words. I looked more closely at Simon, at his condition, at his state of being. He was a complete wreck. His hair was frizzy and going in every which direction, both his face and his large, bushy eyebrows were covered in dirt. There was a fleck of dried blood under his nose. Beneath the coating of grime his cheeks were rosy red and he had this wild-looking gleam in his eyes, like a castaway that's just heard a boat engine.

"Well, Simon, so here you are at Western States," I affirmed, inspecting his physical dilapidation. "Do you still like me?"

He erupted into this wry jackal-sounding outburst. "Yeah, mate," he exclaimed. "I still like you," he said, cackling on.

"Will you continue?" I questioned.

"Yeah, mate, I'll continue. It won't be fast, it won't be pretty, but I'll get there. I've waited five years. I'll get there."

He said it with such conviction it left little doubt in my mind.

I turned my head down and continued with my task. Once I'd emptied my shoes and retied the laces, I got up to leave.

"Simon, this won't make any sense to you, but I think I signed that book for the both of us. Thank *you* for being here; there is more to our chance encounter today than you will ever know."

He blinked his eyes several times and smiled.

I patted him on the shoulder. "Cherish that buckle, mate." Then I turned and tiptoed back to the main aid station to find Kim.

"All set?" I asked her. "I just got a jolt of inspiration."

"Yep, sure. Let's keep the wheels turnin'," she replied. "Good call."

With newfound purpose we headed back onto the racecourse. If I had influenced even one individual to put on a pair of runners and hit the trail, it was all worthwhile. My world had become a brighter place.

23

THE LIGHT

By mile 90, everyone's a believer.

Kim continued ushering me along through the murky cataracts of dawn, that fuzzy intermediary of not being asleep though neither being fully awake. I'd been running now for twenty-four hours, which was something foreign to me at Western States. The thought of taking more than a day to finish the race was never part of my sphere of consideration in the past. Yet here I was. Kim reminded me that in any given year fewer than 20 percent of the runners finished in less than twenty-four hours, and this year with the heat and humidity that number would surely be even lower. Prior to today, I thought the ratio was the other way around, that most people finished in less than twenty-four hours. But such was not the case. Ironically, however, the race winner, Jim Walmsley, had finished more than half a day ago. The time difference between racers at many running events is measured in seconds or milliseconds; in ultramarathons some-times days or series of days split the field. This was a reality I was coming to know.

At the Quarry Road checkpoint—mile 90.7—the first aura of sunlight was beginning to lighten the sky. The past 5 miles had been a demoralizing suckfest. My disposition was now that of a battered survivalist struggling to reach base camp before the storm. I wanted my body to move faster, but it wouldn't, like the recurring nightmare of trying to escape danger on dysfunctional legs.

I grabbed a handful of nuts from the aid station food table and put a few in my mouth. I tried chewing but the desiccated chunks simply migrated around without grounding down. Eventually I gave up and swallowed a mouthful of nut pellets. Kim handed me a glass of water to wash them down.

It was clear that sufficient bodily damage had been done get-ting to this point and that reaching the finish line would require some herculean act of raw fortitude. In absolute terms the 9.3 miles of difficult terrain ahead wouldn't be insurmountable, but after twenty-four hours of merciless pummeling it was an utterly daunting proposition. I made the sign of the cross on my chest and looked skyward. By mile 90, everyone's a believer.

We left the aid station and I hobbled along in a state of wounded destitution, willing myself forward with everything I had. People sometimes ask what lessons I've learned from fail-ure. Failure teaches us many things, mainly that failure really sucks. My unrelenting drive to reach the finish was fueled per-haps less by the aspiration to succeed than the burning desire not to fail. Of course, if I could make it to the finish, both would be accomplished. And that thought compelled me forward.

It is moments like these that define us. Running an ultra obliterates you because few undertakings in life expose people to themselves with such blunt exactitude. We learn what we're made of, we see the true constitution of our character. Here we have our war, and here we learn whether we are cowards or heroes.

This wasn't the race I wanted, not the performance I'd been hoping for. Of course, I hated that. No one wants to fall short of his expectations, and it stung. Though what defines us as individuals is not that we encounter hardship—for everyone does—but how we carry ourselves in such instances when things don't go our way. "There is timing in the whole life of the warrior, in his thriving and declining, in his harmony and discord," the samurai poet Musashi Miyamoto wrote. Like the tides washing in and flowing out, no glory is everlasting. What matters most is having the courage to carry on, fully aware of the tides through which we move, flourishing with dignity and receding with honor, forever dutiful to being all that we can be throughout the cycle.

The pain I felt at this point was eternal, bottomless. Every cell in my body screamed in agony. We'd been pushing hard for the past two hours and I trudged along in a cataleptic state, my senses dulled and narrowed, registering little other than a throbbing electrical jolt that was delivered with each advancing footfall. Sometimes during an ultra, pain is the only indicator that you're still alive.

"How ya doing, Karno?" Kim asked.

Her voice startled me. We were on the final steep climb of the course, the morning sun freshly arisen, the land coming awake.

"I'm hangin' on," I told her, my mouth dry and gritty. I ran my tongue across the front of my teeth and they felt like sandpaper. "Though my pearly whites are now adorned in wool sweaters."

Kim reached into her pack. "Here, take this."

She handed me a stick of gum and I stuck it in my mouth and

started to chew. It brought otherworldly delight. Low-level pleasures take on outsize proportions during an ultra. That stick of gum was pure ecstasy.

"See that light up there?" Kim said, pointing to the faint glow of a lantern whose dim illumination was being overpowered by the intensifying morning sunlight. "That's Robie Point."

Robie Point marks the terminus of the trial section of Western States, after which the remaining course is on the backstreets of Auburn, until you reach the finish line on the Placer High School track.

It's been said of many things that it isn't over until it's over, but at Western States those words couldn't ring truer. Racers have tapped out with fewer than 2 miles remaining; on one occasion the race leader crumbled to the ground within clear view of the finishing tape, unable to continue under his own power. I knew these things, yet I still pushed relentlessly. If collapse was to be my fate, so be it. I wasn't willing to leave anything on the course. And that final climb was a grunt, consuming any residual energy from my internal reservoir, one that had long since run dry. There was nothing on the line here other than a finisher buckle, but I still executed every step with unwavering precision and focused determination to the best of my God-given ability. In my mind, anything less would be a sin.

When we exited through the unlocked gate that separated the dirt trail from the paved roadway, the Robie Point aid station came into view. Standing waiting for us were Nicholas and my mom and dad.

When I saw them, my soul temporarily melted, the emotional avalanche buckling my knees. A wellspring of feelings overflowed within me, salty rivulets of tears streaming down both cheeks. I tried to speak through quivering lips but no words came forth. I wiped the tears with the back of my hands, accomplishing little other than smearing dirt across my cheeks. Here was a grown man, fit and able, a man who had lived a full and complete life, standing on the roadside blubbering nakedly—soul laid bare—while dozens of onlookers gazed in silence. I didn't care, couldn't care; the moment owned me.

At this point twenty-four years ago I emerged from the trail in the middle of the night, battered and beaten, unable to continue only to have my father tell me that I must continue, even if that meant crawling, that I could not stop, that I mustn't ever give up the fight. Today he told me that he was proud of me, and I knew that meant not so much for my current accomplishment but for my perseverance over the decades in remaining true to the man I am. I was doing what I was put on earth to do, and in that there is a certain genuineness and purity. For this my father was proud. It had been a long and oftentimes lonely run, and I was still standing.

"Hey, Pops," Nicholas said, "can I run with you to the finish?"

His request brought immeasurable delight to my heart. "Sure, son," I said, again choking back the sobs. "Let's go before I drive these poor people to drink."

As Nicholas, Kim, and I made our departure, the onlookers began to clap. They said kind words and they patted me on the back as we ran past. It was an endearing moment that will stay with me for the rest of my days.

It was a Sunday morning as we made our way along the streets of Auburn en route to the finish. The neighborhood was coming to life and some early morning revelry was starting to stir, like the beginnings of a tailgate party before a big game. Coolers and lounge chairs were being set up on front lawns, noisemakers emerged from storage, and squirt guns and water balloons were being loaded from yard hoses. The "Golden Hour" was approaching.

I was passing through on the early side of the day. In 1985, John Medinger had been pacing Tony Rossmann in this neighborhood section of Auburn when a newspaper delivery boy rode by on his bike. He tossed a paper on the driveway of the house in front of them and as the pair ran past they glanced at the headlines. On the front page was a story of Jim King's win at Western States. Tony turned to John and in an incredulous voice muttered, "You know it's a long race when you're reading about who won in the newspaper while you're still running the damn thing!"

After a little more than a mile of roadway the bleachers of the Placer High School stadium came into view. The three of us made our passage through the entranceway and ran onto the crimson-colored track. As we did, the voice of John Medinger, who was now the finish announcer, came over the loudspeakers:

"Entering the track is Dean Karnazes. He has eleven Western States silver buckles to his credit and now he has a shiny bronze one to add to his collection. It's not getting any easier, is it Karno? Nice work." Yes, John and I were friends. And yes, John is a jokester.

The Placer High School track was a familiar scene, though now it seemed something different, something new. As T. S. Eliot had written in *Four Quartets*, "at the end of all our exploring we will arrive where we started, and know the place for the first time." As we ran along that track, I saw it all through fresh eyes. And it was something grand, something miraculous, something new.

Coming around the final bend and seeing the finish line, an odd sense of tranquility came over me. I was one of many athletes here today and had just lived one of many remarkable experiences. Each of us had our moments of anguish and our moments of elation, we each were wounded and we each slayed our dragons, many were relishing victory and many were still struggling to make it. I was part of this grand celebration of life and I felt a forevermore contentment with my place in the universe.

As we crossed the finish line it was Topher who put the medal around my neck, my parents standing behind him on the track, smiling. We snapped a few pictures, exchanged some high fives and hugs, and then cheered on the next racer that had entered the arena.

Eventually we gathered our stuff and walked together toward the exit and the bleachers, our arms wrapped affectionately over each other's shoulders. If death were to befall me at this moment I would have no regrets, for it would be, as the Greeks say, *kalos thanatos*, a beautiful death. I'd become the man I was meant to be. There needn't be more.

We found our place to sit in the bleachers with a nice view of the finish and prepared to watch the Golden Hour. It is a curious phenomenon at Western States that the vast majority of finishers cross the tape in that final hour. It's almost as though they have magnets in their pockets and the gravitational pull of the finish line dials up several notches. It is also notable that five times the number of people come out to cheer on the final finishers as do the race winners. And to their credit, most of the race winners come out as well.

If you were to bear witness to that final hour you would understand the appeal. It was a spectacle of human courage and valor like none other. There were people on the verge of collapse who somehow found the strength and resolve to wave and crack a smile as they staggered unsteadily across the tape. There were bloodied nipples, bloodied knees, bloodied elbows, and bandaged heads. Everyone was covered in dirt. Manouch Shirvanioun ran across the tape in 29:56:44, the 299th finisher.

For three tense minutes the crowd bit their lips, but Western States would not permit another. The clock stuck 11:00 a.m. and the race officially ended.

Afterward we had a bite of food with Mom and Dad in the Mother Ship and then said our farewells. Another Western States had now been added to our priceless collection of memories. And all was good.

As I drove back to the Bay Area with Nicholas, he was uncharacteristically gushy. I couldn't decipher whether it was from being up all night crewing for me or whether the experience had truly moved him. He said affectionate things, which wasn't like him, and he repeatedly thanked me for making him part of the adventure.

"Nicholas," I said between sentimentalities, "there's something I'd like to tell you."

"Yes Dad."

"You're my favorite son."

He looked at me, and then took a swing at my leg with his fist, one knuckle slightly raised to exacerbate the pain. Thankfully I flinched in time to avoid contact. He struck again.

"Nicholas," I yelled, "keep your hands on the wheel!"

He kept swinging and I kept up the gymnastic maneuvering, trying to dodge impact.

"Nicholas! If one of those blows lands you could be fatherless! Stop it!"

He kept jabbing.

"Keep your eyes on the road!"

Begrudgingly, he let it go and we both had a good laugh.

When we arrived home and pulled into the driveway, Julie was waiting.

"Well, hello, boys," she greeted us.

"How did you know when we'd be home?" I questioned.

"There's this app called Find My Friends," she said mockingly.

I waved my hand at her dismissively.

"Let's see the shield big guy."

"How did you know I finished?" I asked.

"Nicholas called me."

I pulled out the buckle and showed her. "I'm so proud of you," she said, embracing me.

"Thank Nicholas, too," I added, "he was a champ."

True to his youthful spirit, after helping unpack the car Nicholas said good-bye to go hang out with some friends. Duties were done and despite a sleepless night crewing for Pops it was now time for some enjoyment. More power to him; the kid earned it.

Julie helped me get inside and get situated. We had some food and I pulled a bottle of champagne from the refrigerator.

"Care to join me?" I asked.

"You know I don't drink."

"I don't, either. Usually."

I popped the cork and poured two glasses. We toasted, and Julie wet her lips. I, on the other hand, had an honest gulp. Then another. Then I moved over to the couch and stretched out. Then I promptly fell fast asleep.

Julie covered my legs with a blanket and took off my shoes. There I would spend the evening, on the couch in deep and contented sleep. And when my eyes opened tomorrow morning, I would be reborn a new man and it would be the first day of the rest of my life.

CONCLUSION

NEW WORLD DISORDER

Runs end, running is forever.

It's been said that all good writing comes from a place of pain. That could also be said of all good running. And the pain I'm referring to in this instance is not so much physical but something of a deeper psychological nature.

After finishing Western States my running and racing continued. There was a newfound energy in my stride and it felt right, fluid, and natural. The distances and durations of my runs mattered less than the pure delight of running itself. In such a state a man is in harmony with the universe, unburdened and pleasantly flowing through space and time with little resistance, frictionless and free. Every runner knows this occasional and magnificent feeling of weightlessness.

As 2020 got under way my race and travel schedule was insanely packed. Just the way I like it. It was going to be an active and exciting year, one of my busiest ever. A race I particularly looked forward to was a local favorite called the Miwok 100K. This year marked its twenty-fifth anniversary and I smiled gleefully when tapping the enter button on the registration form. It's always best in your own backyard, so the saying goes.

Leading up to the Miwok 100K was a 50K trail race in San Luis Obispo, my old college stomping grounds in central California. My plan was to drive down the day before and camp near the start. That morning on my way out the doorway I received a text message from the race director.

"Call me."

I dialed his number, then wedged the phone against my shoulder to close the door behind me, travel bags in hand.

"The race is canceled."

I was stunned. "What? Why?"

"Covid." That's all he said. I kept the phone against my shoulder and listened to the eerie buzzing sound that persisted after he hung up.

Other races soon followed, cancellation notices coming in ominous dribs and drabs, each one seeming more surreal and cataclysmic. *This can't be happening, not in this day of science and technology*, I told myself. Yet it very much was.

It should have come as no surprise that the Miwok 100K would follow suit, but denial can be a blinding force. When the race director's notification popped into my inbox I stood staring at it with numb comprehension, the full gravity of what was happening finally crossing the blood-brain barrier. This was no fleeting inconvenience; this was what all the headlines kept calling it, a global pandemic.

Still, with characteristic optimism we runners kept up appearances, staging virtual races and hosting Zoom finishing parties, a hundred miniature heads bobbing on a computer screen the size of a toaster. It was amusing until the novelty wore off. Then it served as little more than a depressing reminder of how good we once had it.

I longed for the satisfaction of a live race; the virtual variety had lost its luster. As the scheduled date of the Miwok 100K approached, I hatched an idea. I decided to still run the race, just solo and self-supported. I knew the route and I knew how much I needed to feel the thrill of challenging myself on a long and difficult racecourse.

When the morning of the Miwok 100K rolled around I arose before dawn and prepared as usual—filling my water bottles, double-checking my headlamps, and having one strong cup of coffee—the prerace ritual sparking some of that customary exhilaration.

Julie drove me to the starting line. A notoriously conservative driver, she proceeded at speeds considerably below the limit as my nerves frayed. This, even though there wasn't a soul on the road.

"If you keep driving at this rate I'm going to miss the start," I blurted instinctively.

It took a moment to register, and when the absurdity of my statement sunk in we both broke out in uncontainable hysterics. This went on for quite some time, neither of us able to reclaim composure.

Her eyes still tear-filled, Julie jousted, "Would you still like me to drive faster?"

"Hun," I said between laughs, "I wouldn't care if a horse-drawn carriage blew past us right now. We have all the time in the world."

When we arrived at the start it was entirely devoid of human presence; nothing moved in the predawn darkness, the leaves of the trees hanging motionless, as if covered in wax. Leaving the protective warmth of Julie's car I felt an unfamiliar vulnerability, as though I were separating from all of humankind.

"You sure you want to do this?" she asked.

I looked around at the ethereal landscape. "I think I have to."

She smiled and I closed the car door, feeling oddly as an astronaut must when leaving the spacecraft. I was alone, as I came into this world and as I will leave it. My mortality has never been felt more intimately than at that moment.

I hit go on my watch and began running.

At the top of the first climb the sun emerged from the eastern horizon. I took out my phone to capture a picture; I needed to document my endeavor. Widening my legs to steady, I peered through the viewfinder to adjust the focal point when a thought occurred.

"What am I doing?" I said aloud.

I turned off the phone and slipped it back in my pack. I did the same with my watch. I didn't need to document anything today. Today I needed to do only one thing: run.

And run I did. Through valley and forest, along ridges and over rolling, endless hills of golden brown, the narrow path wild and overgrown, earth reclaiming its soil, covering up and healing the scars of mankind. I felt cut off as I ran; yet I felt strangely connected, a timeless unity with all things. Passing through this land as a temporary inhabitant, I found a certain grace and dignity in my movements, a willing surrender to some higher being. Running is not about racing or winning, it is about the human challenge and the exploration of self. Running is a conversation, an education, a revolution, an awakening. We discover who we are through the movement of our bodies, and there are lessons to be learned in running alone much as there are in running large races. But in the end, the experience is about you and the trail. We are not the sum of our achievements but an ongoing story that continues to be told with each day, with each step. Older, wiser, and stronger in many ways, I've learned to love the good races and the bad ones. For I now see that it is not the medals or the buckles that matter, but the fact that you are still showing up, still moving forward, still putting in the effort, still doing your best. In the end, that is what makes the difference. That you stayed the course, that you remained true to the person you are. That you did your best.

I returned home after sunset and sat on the porch, untying my shoes and placing them on the landing. I looked up at the sky and the heavens and then looked back down at my shoes. My run had ended. The uncertainty of the days ahead will continue. The madness of man and the infinite foibles of the human condition will persist even when the pandemic is over. Though through it all I know that when the yearning beckons I can lace up my shoes and disappear into that special place of aloneness to once again see what really matters. Escape and revival are but a heartbeat away. Every runner knows this place; it is our eternal fountainhead. For me, as for every runner, runs end, running is forever . . .

GRATITUDE

To Nicholas, Alexandria, and Julie, you are the bedrock of my existence and I am eternally grateful. I love you. I cherish you. And I always will.

To Carole Bidnick, my literary agent, dear friend, confidant, and sounding board, my life would never be the same had you not taken a chance on me. It's been a wild and wonderful journey and I owe it all to you. Thank you for believing in me.

To Gideon Weil, my editor at HarperCollins, your guidance, wisdom, humility, and support have shaped me as a writer, and as a person. Thank you for the education, on prose and on life.

An enormous thank you to Jan Schillay of the *LIVE with Kelly and Ryan* show (formerly *LIVE with Regis and Kelly*) for making the Run Across America, and the White House visit, a reality. That was an episode worth a rerun, and it was all on account of your vision and vivacity.

To Kim and Topher Gaylord, to say you are *like* family would do our relationship a disservice; you are family. We can pick it up from wherever we last left off, and that is the mark of an enduring bond. May our adventures never end.

In closing, thank you to my fellow runners. There is a hallowed place every runner knows that is sacred and universal. We are kindred spirits, united by an irreducible connectedness that transcends borders and boundaries. There is nothing easy in what we do; running requires discipline, determination, and resolve. Bravo. Bravo to you. I wish you everlasting strength and endurance; long may you run . . .